ACTIVE Biology

MIKE BAILEY, KEITH HIRST AND RAY WATKIN

HODDER AND STOUGHTON
LONDON SYDNEY AUCKLAND TORONTO

Contents

Introduction

By working through the questions in the pupils' book, students should significantly improve the skills which are fostered by a study of Biological Science, and at the same time increase their knowledge and understanding of the subject. Learning results from active engagement in process tasks and problem-solving. The items cover the major biological concepts stipulated in GCSE courses and are designed to stimulate interest and curiosity in pupils as well as to reinforce understanding. Pupils should become confident in applying their knowledge and understanding in unfamiliar contexts and thus have a much more secure ability to handle principles as required in GCSE examinations. There is also particular emphasis on the personal, social, economic, environmental and technological applications required for GCSE.

Using the book

It is expected that some of the questions will be used as a follow-up to practical work in a particular topic. Where the most relevant topic may not be obvious, this is made clear under the 'context' heading for each question. Where specific knowledge is required to answer a particular question this is indicated in comments on each item. Teachers may also wish pupils to practise a particular skill, and the quick reference chart on pages (v) to (xi) will enable them to identify which items access that skill.

The items are not primarily intended to be used as assessment tests, but they may very easily be employed as tools of diagnostic assessment, especially to determine pupils' progress in the acquisition of skills. Some may also be appropriate for assessment of certain of the skills required for the coursework component of experimental skills, especially in the NEA and Northern Ireland Biology schemes which build up a detailed profile of skills.

The items may be used to provide valuable practice in answering GCSE style examination questions. However, because they are designed for use as part of the teaching process, they are less tightly structured than operational examination questions. For example, the items are often open-ended, whereas examination questions tend to be more restrictive, e.g. 'Give two reasons', or 'Name one factor'. It is hoped that teachers will encourage pupils to answer these open-ended questions as fully as their ability allows and to think broadly, giving more than one point if possible. Pupils do need training in this respect, otherwise many will adopt the 'minimalist' approach.

Differentiation

The items vary in difficulty; some are relatively easy and can be tackled by pupils of all abilities. Some may appear daunting at first sight but, once pupils get to grips with them, often prove much less difficult. The ability to master skills does not always correlate well with overall ability, and pupils who are likely to obtain lower GCSE grades may find little difficulty with, for example, observation or graph work. Much depends on the experience and motivation of the pupil. Many of the items have been successfully trialled with pupils who are expected to gain grades D to G in GCSE. The level for each question suggested in the answers section is intended as no more than a guide for teachers, to enable them to make a quick estimate of whether items are likely to be appropriate for a particular group or individual.

Three levels are suggested – easy, moderate or difficult. A + sign indicates that the question should be accessible to pupils at that level, but the outcome may be further extended by more able pupils.

For less able pupils, it is most important that confidence is maintained. It is essential, therefore, that the necessary preparation has been done before a question is tackled. Teachers may wish to provide additional cueing in order to make sure such pupils can get started on a problem. It may be necessary to give extra information or discuss appropriate strategies and solutions, since weaker pupils often have difficulty in recall even though they are able to understand the basic principles. Poor communication skills can be overcome by an oral interpretation of the introductory information. In general, the items have not been simplified for the least able since it is easier for teachers to convert the data and information into a suitable form than it is to upgrade it for more able pupils.

In marking it is important that positive achievement is recognised. Many of the questions can be answered at several different levels. It is better that marking schemes are adjusted for the ability of the pupils concerned, rather than marks being awarded as a fraction of the best possible answer. Thus it may be adequate for weaker pupils to give one or two points, whereas able children should be able to think of five. It is discouraging for the weaker pupil to be allocated one out of five. Communication skills must also be taken into account, and teachers will be able to judge the level of clarity and precision which is acceptable. Answers in this guide are given in 'teacher-speak' and are not necessarily those anticipated from pupils. It is important to encourage pupils to express themselves scientifically and with as much exactness as can be mustered. Where pupils are being asked to make suggestions, teachers must, however, be careful to accept any logical answers within the bounds of a pupil's experience and not be influenced by their own perception of the 'right' answer.

The skills

The skills which are germane to the study of biological science have been classified in various ways by different agencies. There is inevitably some overlap between categories, and often it is difficult to separate clearly the skills involved in a particular activity. Not all the scientific and experimental skills can be appropriately tested by means of the sort of items presented in the pupils' book. For convenience the main skills involved have been grouped into the categories listed below. The main skills required by each item are given with the answer summaries, but inevitably there will be some disagreement over allocation, and indeed different pupils may be bringing to bear different skills according to the nature of their previous experience.

a) Observation Includes describing gross and fine details, selecting relevant features, finding similarities and differences, and identification by matching descriptions or use of keys.

b) Measurement Includes measuring length and area, reading scales, determining actual size from a scale drawing, and systematic counting using a tally or appropriate stratification.

c) Data extraction Includes extracting specific information from written accounts, tables and graphs, with interpolation or extrapolation where necessary.

d) Data handling Includes drawing simple diagrams; tabulation; translating information into flowcharts, food webs or other convenient format; constructing graphs, histograms, bar charts and pie-charts; performing appropriate calculations.

e) Data interpretation Includes recognising patterns such as simple relationships; drawing valid conclusions; formulating generalisations; recognising variability and anomalies; explaining in terms of scientific theories, models, etc.

f) Reasoning Includes making predictions and inferences; suggesting explanations of unfamiliar facts and observations; formulating hypotheses which are capable of experimental testing.

g) Application Includes solving qualitative and quantitative problems using observations or principles; evaluating applications and their implications; making decisions based on evidence. This category involves those applications which have particular personal, medical, social, environmental, economic or technological significance.

h) Experimental procedure Includes understanding of principles of experimentation such as safety, use of basic laboratory apparatus, use of controls, sampling techniques.

i) Experimental design Includes devising experiments to test hypotheses, identifying variables, specifying apparatus, planning a sequence; indicating measurements and observations to be made; identifying hazards; critical evaluation; suggesting modifications.

Question Number	Title	SKILLS									RELEVANCE				
		Observation	Measurement	Data extraction	Data handling	Data interpretation	Reasoning	Application	Experimental procedure	Experimental design	Personal (health, medical)	Social	Environmental	Economic	Technological
UNIT 1■															
1	Habitats			b		c,d	a								
2	On the playing field			a,b	c,d	e	f								
3	London and Delhi				a	b,c									
4	On the moors				a	b–d									
5	Life in a stream			a		b,c									
6	In the woods				a	b	c,d								
7	Trees and la Mon-tagne Noir			a–d		f,g	e							e	
UNIT 2■															
8	Single-celled animals	a,b													
9	Vertebrate classifica-tion	b,c	d		d		a								
10	Stonefly nymphs	a,b f,g	c,d	e	c,d										
11	A conifer key	a,b						b							
12	Flying	a,b					a,c								
13	Adaptation	all					all								
14	The naked ape			a	d	b,c	e					c–e			
UNIT 3■															
15	Wytham Wood			a–i			j								
16	Pyramids of numbers				all										

The letters a–j within the body of this table refer to question parts.

KEY: ■Relevant to attainment target 2 in the national curriculum

▲Relevant to attainment target 3 in the national curriculum

★Relevant to attainment target 4 in the national curriculum

Question Number	Title	SKILLS									RELEVANCE				
		Observation	Measurement	Data extraction	Data handling	Data interpretation	Reasoning	Application	Experimental procedure	Experimental design	Personal (health, medical)	Social	Environmental	Economic	Technological
17	Pyramids of biomass and energy				b				a	c					
18	Ecosys-tems and energy				a–c e–g j	h	d,i								
19	In the desert			a,c	b	c	d								
20	An oak wood			b	a	c									
21	Discover-ing feeding relation-ships					b–d				a					
22	Predator–prey re-lationships				a	b–f a,c									
23	Populations			b	e	f,g	d								
UNIT 4■															
24	The water cycle				a	d	b,c	c,d					b		c,d
25	The carbon cycle			a,b	a,b,d			c,d					c,d		
26	The nitro-gen cycle in a meadow			a,b		c		d							d
27	The nitro-gen cycle budget			a	c	b	d						c,d		
28	Nitrogen fixation in root nodules	a,b				c,d	e–h								
29	Fungal symbiosis					a,b	c								
30	A lawn problem			a,b			c,d								all

The letters a–j within the body of this table refer to question parts.

KEY: ■Relevant to attainment target 2 in the national curriculum
▲Relevant to attainment target 3 in the national curriculum
★Relevant to attainment target 4 in the national curriculum

Question Number	Title	SKILLS: Observation	Measurement	Data extraction	Data handling	Data interpretation	Reasoning	Application	Experimental procedure	Experimental design	RELEVANCE: Personal (health, medical)	Social	Environmental	Economic	Technological
31	The compost heap				e–g				a–d						
32	New House Farm			a	b,c			d–f							all
UNIT 5■															
33	Fertiliser pollution			a,b			c,d						all		
34	Pollution indicators					a	b			c			all		
35	Whale hunting				a	b	c							c	
36	Culling							all						all	
37	Air pollution				a,b			d					all		
38	Lead poisoning			a–c	a–c		d				all		all		
39	DDT				b,c	d	a,e	g		f			all		g
40	Biological control			a,d	d	c,e	b	f–j						c–j	c–j
UNIT 6▲															
41	Cells	a,b, d,g,h	c		g,h	d	c,f								
42	Water			a	b,f	c	e,f	d,g,h		h	f			h	h
43	What are we made of?				a	b–c									
44	Food tests					all									
45	Effects of temperature on an enzyme			b	a	c–e			f,h	g					
46	New and old spuds					b–d	e		a,c						
47	Cheese-making			b	a	c		d							c,d
48	Brown potatoes					a,b,e	h–j c,f	h–j	c	f,g					h–j

The letters a–j within the body of this table refer to question parts.
KEY: ■Relevant to attainment target 2 in the national curriculum
▲Relevant to attainment target 3 in the national curriculum
★Relevant to attainment target 4 in the national curriculum

Question Number	Title	SKILLS										RELEVANCE			
		Observation	Measurement	Data extraction	Data handling	Data interpretation	Reasoning	Application	Experimental procedure	Experimental design	Personal (health, medical)	Social	Environmental	Economic	Technological
UNIT 7▲															
49	Leaves	d	f		c,d	c	e,g		a,b						
50	Temperature and photosynthesis		a,b		c		d,e				d	d,e			
51	Stomata	a–c		a–c		d									
52	How do stomata open and close?		a,b		a,b	c	d								
53	Oxygen production photosynthesis	a				b,c		d		d					d
54	Light intensity and photosynthesis				h	g,i,j			a–f						
55	Can we improve on nature?			a,b		c–e		f							e,f
56	The 'Green Revolution'			a–c				d				d		d	all
57	Food crops			a–c		d	f	e,g				e–g		f,g	e
UNIT 8▲															
58	Energy from sweets				a–e		f,h				all				
59	Milk			a–e f–i						all					
60	Square meals			a,d	b	c,d	e–g				all				
61	A hole in the stomach			a		a,d	f	g	b,c,e	f		g			
62	On the farm			a,b,e		f	d	c,g						c,g	all

The letters a–j within the body of this table refer to question parts.
KEY: ■Relevant to attainment target 2 in the national curriculum
▲Relevant to attainment target 3 in the national curriculum
*Relevant to attainment target 4 in the national curriculum

		SKILLS										RELEVANCE			
Question Number	Title	Observation	Measurement	Data extraction	Data handling	Data interpretation	Reasoning	Application	Experimental procedure	Experimental design	Personal (health, medical)	Social	Environmental	Economic	Technological
63	Digestive enzymes					a–c d–f									
64	The dangers of a sweet tooth			a,b		c,e	d				all				
65	Two toothpastes	a	b				c	d,e			all				e
66	The fluoride problem			b,c		a,d		d–g			all	e,f			g
UNIT 9▲															
67	Diffusion				a,b	c	d								
68	Intravenous feeding		a		a	b	d	d	c		d				c
69	Counting red blood cells	b	b		c,d			e	a		e				
70	Athletes' hearts				a	b,d		c			c				
71	A baby's blood system					a–c		d			d				
72	Heart disease						c	b–e			all				
73	Water loss				a		d–f	h	b,c						h
74	Prize pears						b,c			a					all
75	Mineral ion transport			c,d		e			a,b						
UNIT 10▲															
76	Exercise				a	b,c	d	c			all				
77	Breathing					a,b	c,d								
78	Lung structure		a,e		b,c		h,i		c		d,h,i				

The letters a–j within the body of this table refer to question parts.

KEY: ■Relevant to attainment target 2 in the national curriculum
▲Relevant to attainment target 3 in the national curriculum
★Relevant to attainment target 4 in the national curriculum

Question Number	Title	SKILLS										RELEVANCE			
		Observation	Measurement	Data extraction	Data handling	Data interpretation	Reasoning	Application	Experimental procedure	Experimental design	Personal (health, medical)	Social	Environmental	Economic	Technological
79	Lungs and size				d	a,b	c–e								
80	Walking and running			a		b,d	c,e								
81	Running, swimming and flying			a,b	c,d		e,f,g,i	h							
82	Diving ducks			a	b	d	c,d								
83	Hiberna-tion			a,b			c–i								
84	Wine				b–e	a	f–h	f–h							all
UNIT 11■▲															
85	Camels in the desert					a	b,c	b,c							
86	A desert plant						a,b								
87	Spiny ant-eaters					a	a,b								
88	Kidney machines						a–c	d,e			all				e
89	A crab's problem					d	e								
90	Diabetics			b,c		d		e,f	a		all				
91	An aquatic toad				a	b,c									
UNIT 12▲															
92	Reaction times			c	a,b d–f						all	f			
93	Eyes					b,c,e	a,d	d,f		g	all				
94	The thy-roid gland					a–c	e,f		d		all	f			d,f
95	Photo-tropism					a,b	c			d					

The letters a–j within the body of this table refer to question parts.

KEY: ■Relevant to attainment target 2 in the national curriculum
▲Relevant to attainment target 3 in the national curriculum
★Relevant to attainment target 4 in the national curriculum

Question Number	Title	SKILLS									RELEVANCE				
		Observation	Measurement	Data extraction	Data handling	Data interpretation	Reasoning	Application	Experimental procedure	Experimental design	Personal (health, medical)	Social	Environmental	Economic	Technological
96	Geo-tropism						b			d					
97	Stickle-backs			b		a,c,d									
98	Waterfleas						a,b								
UNIT 13▲															
99	Growing up			a,b	c	e–f	d				f,g,h				
100	Ageing			b	a	c	d	c			all	e			
101	Cell division	a,d			b	c									
102	Flowers	a,e				b,e,f	d								
103	Human cir-culation	a,c,d				b	b				b				
104	Population growth					a–c	a–c					all			
105	Contracep-tion			a–c		d		e			all	e			
106	Gestation			a	a	b,d,e	c				e				
107	Eggs						c–e								
UNIT 14★															
108	Finger-prints	a,b,d				e	c				all				
109	Tobacco plants	a			b	c–e	f								
110	Maize	b			d	a,d									
111	Oil palms					a,d	b	c,e						c,e	all
112	Blood groups			e	a,b,d,e	c	f			all					
113	Ladybirds			a–d g		e,f, c,j	g,h,k								
114	The pep-pered moth			c		a,b,d	e		g				b,e		
115	Wheat				a,g	b,c	d,f,i	c,j						e	all
116	Adaptation					all									

The letters a–j within the body of this table refer to question parts.

KEY: ■Relevant to attainment target 2 in the national curriculum
▲Relevant to attainment target 3 in the national curriculum
★Relevant to attainment target 4 in the national curriculum

UNIT 1 Organisms and their habitats

1 Habitats

Context

The concepts of species being restricted to particular habitats and of environmental factors affecting organisms.

Knowledge required

a Habitat. Some knowledge of common organisms. Reference books might be useful.

b Environmental factor.

Skills

a Knowledge and reasoning (inference)

b Data extraction

c, d Data interpretation

Level

a Easy **b** Mod **c** Mod **d** Mod

Answers

a bluebell – woodland floor
crab – seashore
daisy – grassy field
earthworm – garden soil
heather – moorland
lichen – stone wall
trout – mountain stream
water boatman – freshwater pond

b (i) C (ii) B (iii) A (iv) B (v) C and F

c (i) D (ii) F (iii) A (iv) E (v) B (vi) C

d X – Habitat B (desert)
Y – Habitat E (large river)
Z – Habitat C (inside the intestine)

2 On the playing field

Context

Investigation of a habitat.

Knowledge required

Useful to have practical experience of the technique involved, i.e. using a quadrat.

Skills

a, b Data extraction

c Data handling (tabulation)

d Data handling (calculation)

e Data interpretation

f Reasoning (hypotheses)

Level

a Easy **b** Easy **c** Mod **d** Mod **e** Mod
f Mod+

Answers

a Daisy, plantain.

b Buttercup, clover, dandelion. (Grass should not be included, as the species on the playing field is not identified and cannot be said with certainty to occur in the rough grass area.)

c Number of different types of plant

Quadrat	Playing field	Rough grass
1	3	4
2	3	5
3	3	6
4	4	5
5	4	7

Playing field = 3.4 types/square metre.
Rough grass area = 5.4 types/square metre.

e There are fewer types of plants on the playing field than in the rough grass area.

f Any sensible suggestions, e.g. the number of plants that can live on the playing fields is reduced by: mowing; trampling; compacted soil; use of weedkillers.

3 London and Delhi

Context

Organisms are confined to certain environments by environmental factors such as temperature.

Knowledge required

None.

Skills

a Data handling (graphs)

b, c Data interpretation

Level

a Easy **b** Easy+ **c** Easy+

Answers

a Graph – (top right).
Graph should have:
⋆a heading
⋆sensible choice of scales;
⋆axis headings and units;
⋆accurate plots with points clearly visible;
⋆points linked, preferably by straight lines as intermediate points have no validity;
⋆a key.

b London – steady rise from February to a peak in July/August; followed by steady fall to winter means.
Delhi – steady rise from January to peak in May/June; slight reduction in July/August/September; then steady fall back to December and January lows.

Mean monthly temperatures in London and Delhi

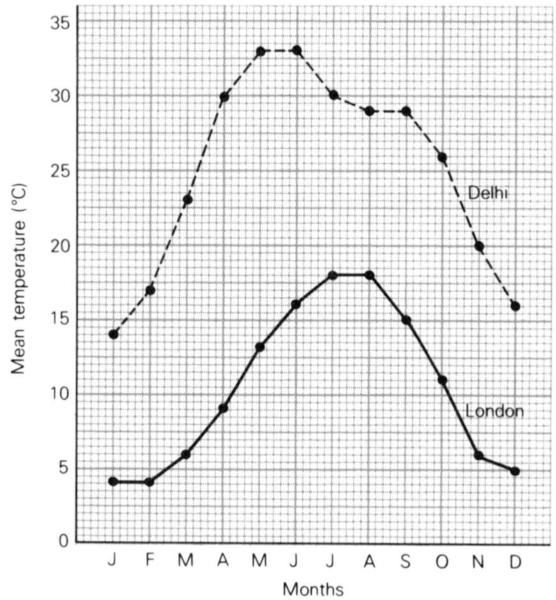

Overall higher temperatures in Delhi and earlier peak may also be noted.

c Firebrat – Delhi (from the data: one type does live in warm bakeries, etc., in the UK).
Housefly – either (but only breeding in London in summer months).
Springtail – either.

4 On the moors

Context

Distribution of organisms in a habitat is affected by environmental factors.

Knowledge required

pH scale.

Skills

a Data handling (graphs)

b to **d** Data interpretation

Level

a Easy+ **b** Easy **c** Easy **d** Diff

Answers

a

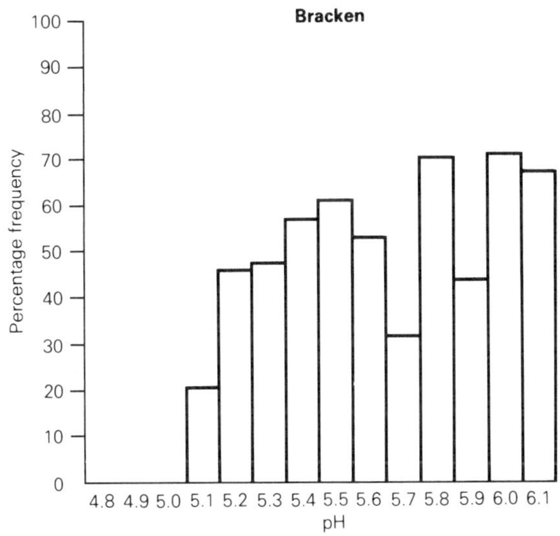

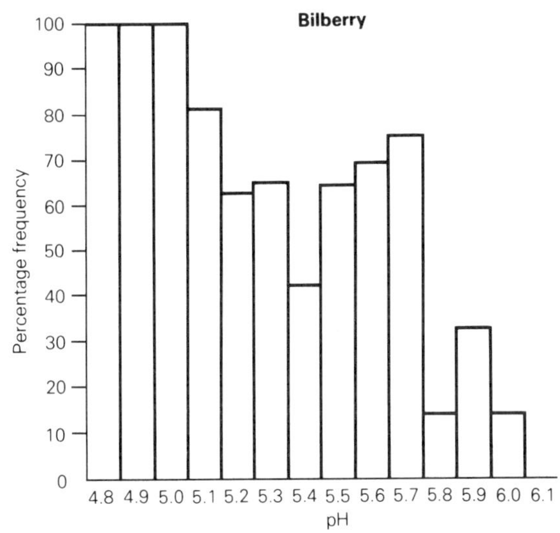

b Bilberry.

c Bracken – found in abundance at pH 6.0 and 6.1, whereas bilberry is very low or absent at these pHs.

d (i) The data shows the percentage frequency of the plants in all the samples (i.e. quadrats) taken at each pH. Some of the samples would contain both species.
(ii) Some samples would have neither bracken nor bilberry.

5 Life in a stream

Context

The distribution of organisms depends on their ability to cope with environmental conditions. This question also introduces the idea of testing organisms under laboratory conditions.

Knowledge required

None.

Skills

a Data extraction

b, c Data interpretation

Level

a Mod **b** Mod **c** Diff

Answers

a (i) 4 (ii) 3

b Mayfly larva. The data states that a current speed of more than 240 cm/s was needed to wash it away, which suggests that this was the maximum current speed tested.

c Maximum current speed the animal moved against. The figures are closer to the maximum current speeds found in the animal's habitat. It would clearly be necessary for the animal to be able to move around at times, and this would determine the animal's ability to survive in a particular stream.

6 In the woods

Context

Interaction between organisms in an ecosystem. One type of organism can change the environmental conditions and thus affect other organisms growing near it.

Knowledge required

b Trees may be either evergreen or deciduous.

c, d Light affects the growth of plants.

Skills

a Data handing (graphs)

b Data interpretation

c, d Reasoning (prediction)

Level

a Easy+ **b** Easy+ **c** Mod+ **d** Mod+

Answers

a

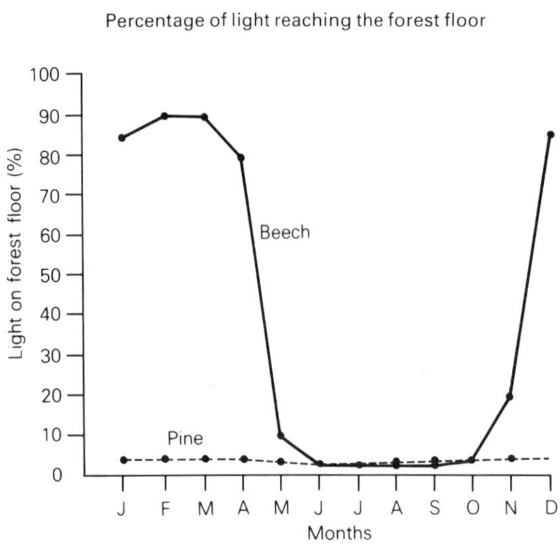

b (i) Pine trees are evergreen and grow close together so that the forest floor is always in deep shade.

(ii) Beech trees are deciduous. In winter months most of the light reaches the forest floor as there are no leaves. In April new leaves grow and deep shade quickly develops and persists through the summer months. In October leaves fall and light penetration increases again.

c Pine trees cut out most of the light, so only shade-tolerant plants can grow.

Beech trees only cut out light in summer, so plants may grow during spring and early summer. Thus plants adapted to have their main growing period at these times are able to survive in beech woods.

d Beech forest is likely to have the greater number of species. Plants such as dog's mercury and bluebell which grow and flower in early spring will be able to colonise the forest floor. Pine forests are likely to have few flowering plants in the denser parts, though mosses, lichens and algae may grow.

7 Trees and la Montagne Noire

Context

Interaction between the organisms and a variety of environmental factors in an ecosystem. The question illustrates some of the complexities involved in determining which of many factors may be involved in the distribution of organisms in an ecosystem.

Knowledge required

e Vines as source of grapes, and hence wine. Access to a suitable tree reference book would help pupils to visualise the trees involved.

Skills

a to **d** Data extraction

e Reasoning (inference)

f, **g** Data interpretation

Level

a Easy **b** Easy **c** Easy **d** Easy
e Easy+ **f** Mod **g** Diff

Answers

a Beech.

b Chestnut.

c Downy oak.

d Grows only by the river.

e (i) Vines produce grapes used to make wine.
(ii) Sensible suggestions include: higher average temperature; grow better on limestone soil; grow better on south facing slope – hence more sunlight; lower rainfall; economic reasons, e.g. near to wine-making facilities.

f (i) The beech grows at and above the 800 m contour with a high annual rainfall around 1300 mm and a mean annual temperature of 11 °C. It grows on a sandstone soil.
(ii) The holm oak grows up to about the 600 m contour in warm (mean annual temperature 12 to 13 °C), dryish (annual rainfall less than 800 mm) conditions. It grows on a limestone soil.

g (i) The holm oak grows only on the limestone, whereas sessile oak grows only on the sandstone, even though they grow at the same altitude. The chestnut is also restricted to the sandstone soils, although conditions would be otherwise similar on the other side of the river. Able pupils may point out, however, that rainfall and temperature differences complicate the picture.
(ii) Only the beech grows above 800 m. The sessile oak can tolerate high rainfall, but does not grow above 800 m. The downy oak appears to be restricted to about 650 to 800 m. However, altitude in itself is unlikely to be the relevant factor. Such associated factors as average temperature or rainfall are likely to be the major influence.
(iii) The downy oak is restricted to the southern side of la Montagne Noire, where rainfall is lower.
(iv) There is no firm evidence from the data to support temperature as the main factor affecting distribution, although it may well be involved, e.g. in the distribution of the vines and holm oak.

UNIT 2 The variety of living organisms

8 Single-celled animals

Context

Diversity and classification. It is not expected that pupils will be familiar with the single-celled organisms shown in the diagrams, nor is it expected that they should learn the taxonomic names of Protozoan classes. Some pupils may, however, be interested to investigate further the variety of single-celled organisms.

Knowledge required

None.

Skills

a, **b** Observation

Level

a Easy+ **b** Easy+

Answers

a There is a variety of acceptable solutions, involving consideration of, for example, the projections, cell shape or cell contents. Thus one possibility is:

A: 1,5,8
B: 2,4,7
C: 3,6
D: 9,10

b A: One hair-like structure (flagellum) projecting from the cell
B: Many small hairs (cilia) projecting
C: A small number of long hair-like projections
D: No projecting hairs

The organisms shown belong to the following genera:

1. *Euglena*	2. *Trachelocerca*
3. *Actinophrys*	4. *Paramecium*
5. *Strombomonas*	6. *Vampyrella*
7. *Stentor*	8. *Peranema*
9. *Amoeba*	10. *Endamoeba*

9 Vertebrate classification

Context

Diversity and classification.

Knowledge required

a The five vertebrate classes.

Skills

a Reasoning (inference)

b, **c** Observation

d Measurement and Data handling (calculation)

Level

a Easy **b** Easy+ **c** Easy+ **d** Mod

Answers

a Fish.

b Streamlined shape; eyes; mouth; presence of fins; position of fins, e.g. single dorsal fin, tail fin, fin on under surface; 'notched' tail fin; gill openings; lateral line.

c *Shark*

Five gill openings
Mouth on underside of head
Fins have broad attachments to body

Fins have smooth surface

Trout
Single opening for gills
Mouth at end
Fins have narrow attachments
Fins have bars (rays)

d Shark 9.0 m
Trout 23.2 cm

10 Stonefly nymphs

Context

Diversity.

Knowledge required

None.

Skills

a, **b** Observation

c, **d** Measurement and Data handling (calculation)

e Data extraction

f, **g** Observation

Level

a Easy **b** Easy **c** Mod **d** Mod **e** Easy
f Easy+ **g** Easy+

Answers

a C is the same as species X.

b The shape of the head, the shape of the tails (cerci) and the thoracic segments are all possible answers.

c Approximately ×5.8 (between ×5.6 and ×6.0 would be reasonable answers)

d Allow ± 1 mm in measuring body length.

Nymph	Magnification	Length of body on diagram/mm	Actual body length/mm
A	×4	37	9.25
B	×4	40	10
C	×4	32	8
D	×4	43	10.75
E	×3	36	12
F	×1.5	32	21.3
G	×2	29	14.5
H	×5	41	8.2
I	×4	42	10.5

e Smallest – H; largest – F.

f 6 legs; 2 antennae; 2 tails (cerci); legs jointed; claws on legs; bristly edge on legs; segmented body; 3 segments in thorax; 2 small projections on head. (Names of parts are not required; suitable descriptions should be acceptable.)

g

Nymph F	Nymph G
Antennae straight	Antennae curved
Head squarish	Head broad and short
Thorax narrow	Thorax broad
No structures between legs	Gill-like projections on thorax between legs
10 segments on abdomen	8 segments on abdomen
Narrow joints between abdomen segments	Wide joints between abdomen segments
Cerci 'hairy'	Cerci not 'hairy'
Body about 21 mm long	Body about 14.5 mm long

11 A conifer key

Context

Use of keys. The National Criteria for Biology GCSE require pupils to be able to use simple keys based upon observable features. This key is slightly more difficult than those usually encountered in examination questions, as it includes a number of species not illustrated. On the other hand, this makes the use of the key more realistic. The construction of a key is not required for GCSE, but might be used to extend more able pupils and help them to understand the principles on which a key is based.

Knowledge required

How to use a dichotomous key.

Skills

a Observation

b Observation and Application

Level

a Mod **b** Diff

Answers

a

Species	Name of conifer
A	European Larch
B	Giant Fir
C	Western Hemlock
D	Scots Pine
E	Cedar of Lebanon

b Pupils will devise a number of valid keys. Descriptors should be free-standing, that is, not dependent on comparison with another specimen, e.g. 'longer' or 'wider'.

One possible key is:

1. (a) Seed more than 25 mm long — Go to 2
   (b) Seed less than 25 mm long — Go to 3
2. (a) Wing narrow and coming to a point — **Scots Pine** seed
   (b) Wing broad and flat at the end — **Giant Fir** seed
3. (a) Wing longer than it is wide — Go to 4
   (b) Wing nearly as wide as long — **Japanese Larch** seed
4. (a) Wing with pointed end — **European Larch** seed
   (b) Wing with rounded end — **Sitka Spruce** seed

12 Flying

Context

Diversity and adaptation. The questions require pupils to consider how wings on a variety of different organisms are adapted for their function.

Knowledge required

None.

Skills

a Observation and Reasoning (inference)

b Observation

c Observation and Reasoning (inference)

Level

a Mod **b** Mod **c** Diff

Answers

a (i) Any two of: large surface area; thin; relatively long and narrow; contain supporting framework

(ii) *Large surface area* to exert force against air.
Thin to 'cut through air', and to reduce overall mass
Long and narrow to provide correct aerodynamic shape and reduce drag, or any

reasonable suggestion. (There is no simple answer, but fast-flying birds like swifts tend to have longer, narrower wings.)
Framework to keep wing out-stretched. No detailed knowledge of the mechanism of flight can be expected. Interested pupils may wish to find out more in a suitable biology (or bird) book.

b (i)

Dragonfly
Two wings
Rounded end
Attached to body by narrow joint
Main support near front of wing

Flying fish
Single wing or fin
Pointed end
Has broad attachment
Ray-like supports throughout fin

(ii)

Bat
Skinny membrane
Bony support visible
Wing attached to both front and rear limbs
Small claw projects from wing

Falcon
Covered in feathers
No internal support visible
Wing attached to front limb only, or only one limb visible
No projecting claw

c Possible suggestions include: flatter structure; less flexible – no need to bend during upstroke; less strong muscle and attachments.

13 Adaptation

Context

Diversity and adaptation. The problem requires pupils to consider conditions in a variety of different habitats, and the special features which adapt certain plants to live in these conditions. There is no clear 'right' answer, but pupils should be encouraged to look for reasons which allow organisms to occur in one set of conditions but not another.

Knowledge required

Adaptation. Reference books would help pupils to find out more about the features of the organisms given.

Skills

Observation and Reasoning (inference)

Level

Mod+

Answers

Organism	Set of conditions	Special adaptation for conditions
Water lily	E	Large, round, flat leaves (air-filled)
Greater plantain	H	Flat ring of leaves; 'tough' leaves
Lichen	A	Able to survive drying out
Toadstool	C	Hyphae can digest dead wood
Cactus	D	Able to store water; thick cuticle to reduce water loss
Ivy	F	Adhesive roots
Knotted Wrack	I	Air bladders
Venus flytrap	G	Able to digest insects (as source of nitrogen compounds)
Moss	B	No protection from water loss; tolerates acid conditions
Spruce tree	J	Sloping branches to shed snow; thick cuticle to reduce water loss in cold conditions

14 The naked ape?

Context

Diversity. The question illustrates the point that there are measurable differences between closely related organisms, and even between races of human beings.

Knowledge required

None.

Skills

a Data extraction

b, **c** Data interpretation

d Data handling (calculation)

e Reasoning (hypotheses)

Level

a Easy+ **b** Mod **c** Easy **d** Mod **e** Mod+

Answers

a Apes have hair over the back and chest as well as on the head. Human beings have little or no hair on the back or chest.

b (i) Orang-utan, as it has hair evenly distributed over all three parts.

(ii) Area of each part covered by hair. Average number of hairs on the arms.

c Mongolian.

d 165 000

e Possible suggestions include: insulation from sun's rays, or to reduce heat loss; protection from mechanical damage; sexual attraction.

Some pupils may be interested in pursuing further the question of human lack of hair, and may be directed towards the book *The Naked Ape* by Desmond Morris. In the first chapter several possible explanations for human nakedness are suggested.

UNIT 3 Ecosystems

15 Wytham Wood

Context

Food chains and webs. The food web given is rather more complex than that usually encountered in examination questions but gives a fuller picture of an actual community.

Knowledge required

Understanding of how a food web is constructed is required for all parts.

d to **f** The terms: herbivore, omnivore and carnivore.

g Food chain.

h Primary consumer.

i Trophic levels – producer, primary, secondary and tertiary consumer.

Skills

a to **i** Data extraction

j Reasoning (prediction)

Level

a Easy **b** Easy **c** Easy **d** Easy **e** Easy
f Easy **g** Easy **h** Mod **i** Mod **j** Mod

Answers

a Mice, voles and shrews.

b Leaf litter and fungi.

c Leaf roller moth.

d Any two of: leaf roller moth, winter moth, 'insects', earthworms, soil insects and mites.

e Mice and voles.

f Any two of: weasel, owl, shrews, rove beetles, moles and ground beetles.

g (i) herbs → insects → rove beetles → shrews → owl
(ii) leaf litter → fungi → mites → ground beetles (or rove beetles) → moles

h 9 (including 4 which feed both as primary and secondary consumers).

i Any four of:
mice, vole, soil insects, mites – primary and secondary consumers;
rove beetles, ground beetles, weasel, shrews, moles, owls – secondary and tertiary consumers. (Strictly shrews, moles and owls all feed in longer food chains.)

j (i) Increased feeding damage to oak and other trees and bushes.
(ii) More food available for mice, voles and rove beetles. Hence there would be likely to be more breeding success and larger numbers produced. This in turn would provide more food for the tertiary consumers and again perhaps an increase in numbers.

16 Pyramids of numbers

Context

Communities and pyramids of numbers. This question requires pupils to appreciate that the 'standard' pyramid shape does not always apply, and therefore leads to the idea of the pyramid of biomass as giving a more helpful representation.

Knowledge required

Food chains and webs; producers; pyramids of numbers; trophic (feeding) levels.

Skills

Data handling (translation)

Level

Mod

Answers

a vegetation → water buffalo → mosquitoes E

b grass → wildebeest → lion A

c berries → blackbird → lice → protozoa F

d phytoplankton → zooplankton → herring → cod C

e rose bush → greenfly → ladybirds → blue tit B

f leaf litter → earthworms → hedgehog; leaf litter → beetles → hedgehog; leaf litter → [spiders] → hedgehog D

17 Pyramids of biomass and energy

Context

This question considers the more sophisticated concepts relating to biomass and energy at different trophic levels in a community.

Knowledge required

a, b Biomass and dry mass. Trophic levels.

c Principles of calorimetry.

Skills

a Experimental procedure

b Data handling (graphs)

c Experimental design

Level

a Diff **b** Diff **c** Diff

Answers

a The amount of water inside an organism varies according to the conditions. Dry biomass gives a constant and comparable value.

b Various arrangements are possible, but in each case the diagram should be accurate and to scale. Using 2 mm graph paper and a scale of 1 cm^2 to 25 g/m^2 allows one small square to represent 1 g/m^2. For energy, a scale of 1 cm^2 to 500 kJ/m^2 allows one small square to represent 20 kJ/m^2, which emphasises the similarity between the two diagrams.

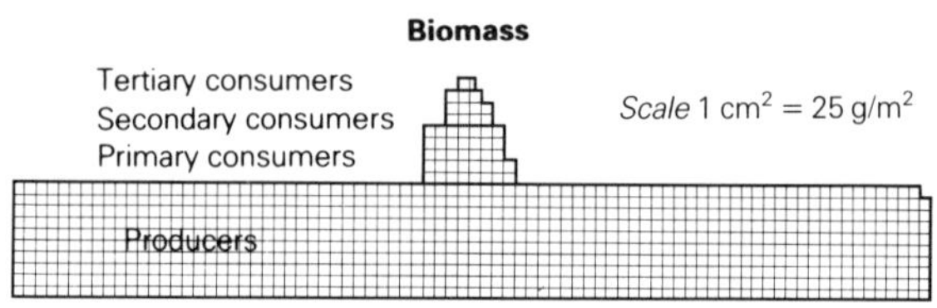

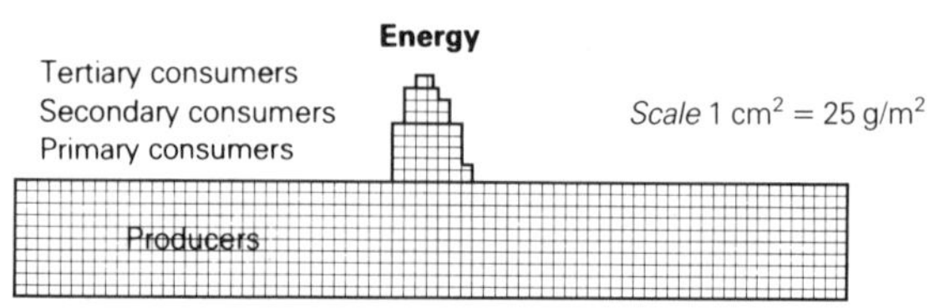

c By use of calorimetry on a representative sample of dried organisms. The experimental sequence might be: collect a suitable sample; kill and oven dry to constant mass; burn a known mass in a calorimeter; measure the temperature rise of the water; calculate the energy value per gram; calculate the total energy content from the biomass information.

18 Ecosystems and energy

Context

This question offers a more detailed consideration of energy flow in an ecosystem. Pupils should appreciate the large amounts of energy 'lost' from each trophic level.

Knowledge required

Photosynthesis, respiration.

Skills

a to **c** Data handling (calculation)

d Reasoning (prediction)

e to **g** Data handling (calculation)

h Data interpretation

i Reasoning (explanation)

j Data handling (graphs)

Level

a Mod **b** Mod **c** Mod **d** Diff **e** Mod
f Mod **g** Mod **h** Mod **i** Mod **j** Mod

Answers

a $\frac{50\,000}{4\,000\,000} \times 100\% = 1.25\%$

b $50\,000 - 39\,000 = 11\,000$ kJ per m^2 per year

c $\frac{11\,000}{50\,000} \times 100\% = 22\%$

d Reflection, or absorption and conversion to heat, which is ultimately dissipated through radiation, evaporation of water, etc. Some sunlight may be transmitted through leaves, or gaps between leaves, to the forest floor; some will fall upon non-photosynthetic parts of the trees.

e $22\,000 - 1065 = 20\,935$ kJ

f $\frac{482}{20\,935} \times 100\% = 2.3\%$

g $\frac{53}{20\,935} \times 100\% = 0.25\%$

h Bones, skin and hair would probably be left.

i Maintaining body temperature; synthesis of new materials in growth and repair; movement; other muscular activity, e.g. in chewing, peristalsis, heartbeat; activities of body organs, e.g. brain; effective transport; etc.

j

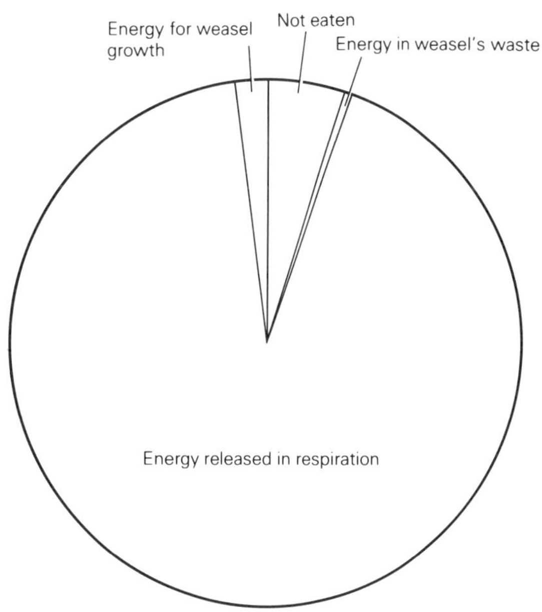

19 In the desert

Context

Communities and food webs. Pupils are required to work out the feeding relationships in an unfamiliar community. Simple 'general' names have been used to describe the organisms involved.

Knowledge required

a Primary, secondary and tertiary consumers.

b Food webs.

c Food chains.

d Idea of pyramid of numbers.

e Energy derived from sunlight via photosynthesis.

Skills

a Data extraction

b Data handling (translation)

c Data extraction

d Reasoning (prediction)

e Data interpretation

Level

a Easy **b** Mod+ **c** Easy **d** Mod+
e Mod+

Answers

a (i) Any three from: dung beetle, ant, bristletail, beetle larva, termite.

(ii) Any three from: antlion larva, spider, worm snake, camel spider, lizard, scorpion.

(iii) Any two from lizard, scorpion, camel spider.

b The layout of the pupils' foodwebs will vary, but they should include the links shown in the example below.

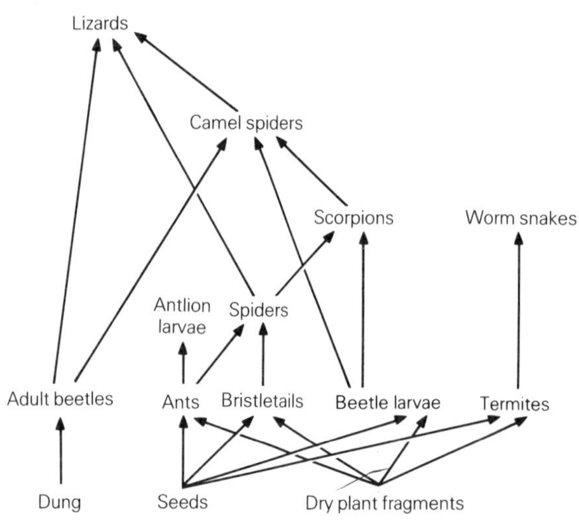

c Several solutions are possible for each of the food chains, e.g.

(i) seeds → ants → antlion larva
(ii) dung → dung beetle → camel spider → lizard
(iii) dry plant fragments → beetle larvae → scorpion → camel spider → lizard
(iv) seeds → bristletail → spider → scorpion → camel spider → lizard

Pupils should be encouraged to include the producer/plant material in the food chains.

d Pupils should appreciate from the pyramid of numbers that primary consumers will be most common. Some may pick termites because they apparently only have one predator, but more able pupils will realise that there is no specific evidence to justify this. Dung beetles are unlikely to be most common, since only a small proportion of available plant matter is likely to be eliminated as dung.

The lizard is likely to be least common, as relatively large animals need a large mass of food and have a low reproductive capacity. There is also a high energy loss at each trophic level, and therefore less energy is available to secondary and tertiary consumers.

e The seeds and plant fragments would have originated from photosynthesising plants, and hence depended on the sun's energy. There would also be a similar origin for the plant material in dung; animal material would ultimately have been derived from plants.

20 An oak wood

Context

Biomass and productivity. The data illustrate the different productivity of the different layers in an oak wood. It might be worth comparing the absolute net production in each layer with the percentage change, and discussing the factors involved.

Knowledge required

Biomass. Net production.

Skills

a Data handling – calculation and tabulation

b Data extraction

c Data interpretation

Level

a Mod **b** Mod **c** Mod+

Answers

a

Region	% increase in biomass
Oak tree canopy	2.5
Shrub layer	11.0
Herb layer	85.7
Fallen leaves	∞
Roots	4.2

b (i) Herb layer (since leaf fall does not represent production).
(ii) Oak tree canopy.

c Credit should be given for any sensibly argued answers showing understanding of possible experimental error. It is, however, most likely that the figures are underestimates, since, for example, they would not take into account material eaten by herbivores, or plants which had grown and died in less than one year.

21 Discovering feeding relationships

Context

The question requires pupils to consider the practical problems involved in finding out what animals feed on, and introduces them to the use of radioactive isotopes as tracers.

Knowledge required

Familiarity with radioactive isotopes would be helpful. The idea of a food chain is required for part **d**.

Skills

a Experimental design

b to **d** Data interpretation

Level

a Mod+ **b** Mod **c** Mod **d** Diff

Answers

a The most likely suggestions are:

(i) to observe the animals feeding, either directly or with film;
(ii) to catch the animals and analyse the gut contents, or to examine faeces or pellets.
Another possibility is to offer a range of possible foods and observe the choices made. Discussion can reveal the practical difficulties of these methods in the natural environment.

b The insect is feeding on the living plant material, since the radioactivity appeared in the insects in the area where the living grass was treated, but not where the detritus was treated.

c The snail is feeding on the detritus, since the radioactivity appeared in the snails in the area where the detritus was treated, but not where the grass was treated.

d (i) The spiders would not be feeding directly on the treated material. They only acquired radioactivity after feeding on organisms which had fed on the treated material.
(ii) The spiders are secondary consumers and must be feeding on either the insects or the snails, or possibly both.

22 Predator–prey relationships

Context

Factors affecting population size. The data show how the numbers of predators and prey are interrelated.

Knowledge required

Carnivore.

Skills

a Data handling (graphs)

b to **f** Data interpretation

Level

a Mod **b** Mod **c** Mod **d** Mod **e** Mod
f Mod

Answers

a These are fairly complex graphs to choose scales for and to plot. It may be necessary to provide suggested scales for less able pupils, and to adjust these so that the two graphs can easily be compared. More able pupils may be guided to draw the two graphs on the same time axis as shown below.

b The peaks for the plant mites occur before the peaks for the carnivorous mites.

c (i) The carnivorous mites eating the plant mites.
(ii) The increase in the carnivorous mite population to a peak of 40 on day 24 as the plant mite population declined.

d The large amount of food available to the carnivorous mites because the plant mite population was so high.

e Probably shortage of food as the plant mite population declined.

f Low carnivorous mite population and ample supply of food for the plant mites.

Graph showing the population changes of a plant mite and its predator, a carnivorous mite over 120 days

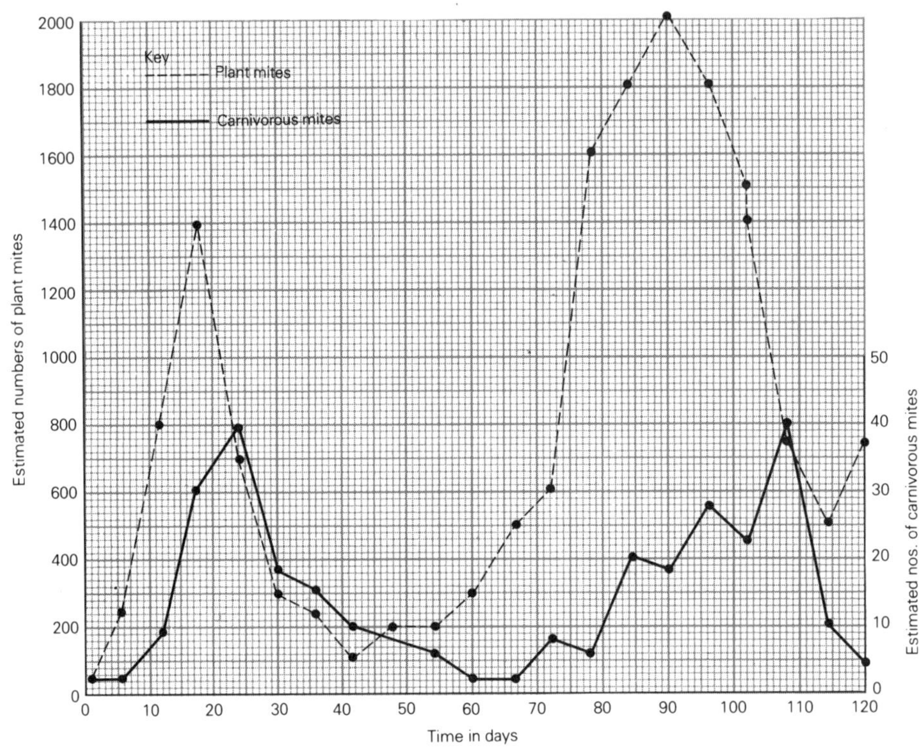

23 Populations

Context

Population growth and factors limiting population size. The example illustrates more or less unrestricted population growth in the early stages of colonisation of a habitat.

Knowledge required

Familiarity with factors affecting population growth would help.

Skills

a Data interpretation

b Data extraction

c Data interpretation

d Reasoning (hypotheses)

e Data handling (graphs)

f, **g** Data interpretation

Level

a Mod **b** Easy **c** Mod **d** Mod **e** Easy
f Mod+ **g** Diff

Answers

a (i) Adult birds breed in spring. The number of young produced is more than the number of adults that die.

(ii) In winter a number of birds die due to food shortage, cold, etc., and no young are being produced to replace them at this time.

b 1941

c The increase between April and September 1941 was about 850. Between April and September 1942 it was only 575.

d Any two suggestions such as shortage of food, shortage of space for breeding, disease.

e

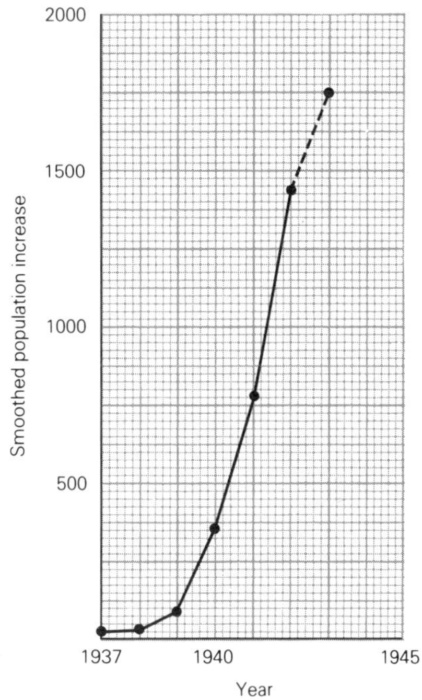

f The rate of increase is slow at first and then increases rapidly, i.e. showing exponential growth. This can be explained by considering the doubling series – i.e. 2, 4, 8, 16, 32, etc.

g The curve might be expected to flatten out again, as limiting factors to population growth take effect.

UNIT 4 Natural cycles

24 The water cycle

Context

Water and its supply. As well as the basic water supply the question considers some of the practical problems of maintaining a water supply. This is therefore an opportunity to address some of the environmental, technological and social issues involved.

Knowledge required

a, **b** Transpiration.

c Waterlogging; osmotic effect of strong solutions.

Skills

a Data handling (translation)

b Reasoning (explanation)

c Reasoning (explanation) and Application

d Data interpretation and Application

Level

a Mod **b** Mod+ **c** Diff **d** Mod+

Answers

a A wide range of diagrams is possible, varying from the pictorial to the schematic. The best responses will include all the information given in the paragraph. A possible solution is shown.

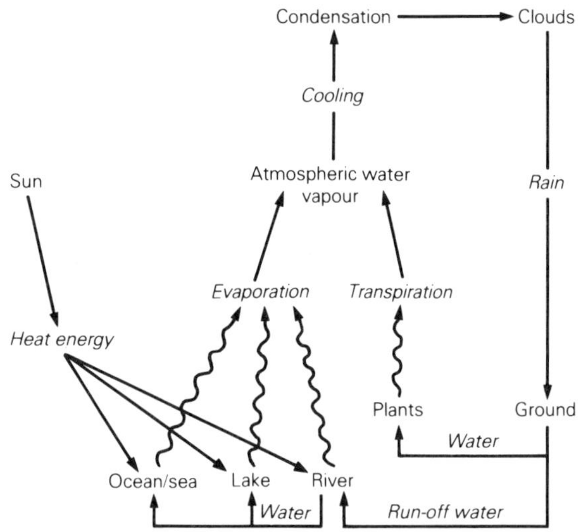

b Deforestation results in there being fewer plants (or a reduced mass of plant and a much smaller surface area) to release water vapour by transpiration. Therefore less cloud is formed and less rain falls, so the climate becomes drier. This in turn may lead to a further reduction in plant growth and problems of erosion, desertification, etc.

c (i) Waterlogging reduces soil oxygen content. Plant roots therefore cannot respire and may die, or be unable to take up sufficient mineral ions by active transport.

The increased salt concentration reduces the water potential and therefore reduces or prevents water uptake. Plant growth is therefore retarded or prevented.

(ii) The water may be drained away by using drainage channels, land drains, etc. Pupils may suggest improving the soil texture by adding sand, lime, etc., although this is unlikely to be practicable on a large scale. To remove the salts a through-flow of water is needed, and this may require pumps, but this is a costly solution.

d (i) The small reservoirs provide more flood storage capability and protect a much larger area from flooding. Therefore there is likely to be less damage to farmland or stock.

There is a smaller area of valley bottom used for the reservoirs, and therefore more agricultural land is preserved.

There is a rather larger area available for recreation. The total cost is less than one third that for a single large reservoir. (It might be worth thinking about the explanation for this.)

(ii) Access roads to 34 separate sites, especially if the reservoirs were used for recreation, could cause greater environmental impact. More staff might be required to monitor the separated reservoirs. Any other sensible suggestion should be accepted.

25 The carbon cycle

Context

The carbon cycle. Pupils are required to consider the processes involved on a global scale.

Knowledge required

c Combustion of fuels.

d The greenhouse effect.

Skills

a, **b** Data extraction and handling (calculation)

c Knowledge and Application

d (i) Data handling (graphs)
(ii) Application and communication

Level

a Mod **b** Mod **c** Easy **d** Mod+

Answers

a Angle subsumed by CO_2 fixation arc = 174°.

Therefore amount = $\frac{166 \times 174}{360}$ = 80 gigatonnes (approx.)

b Angles = decomposers 76°
producers 76°
consumers 17°
TOTAL 169°

Therefore amount = $\frac{166 \times 169}{360}$ = 78 gigatonnes (approx.)

c Burning fossil fuels, e.g. coal, oil, gas, etc.

d (i) Graph. Estimate for year 2000 is about 0.034%.

Graph showing the percentage of CO_2 in the atmosphere between 1900 and 1983

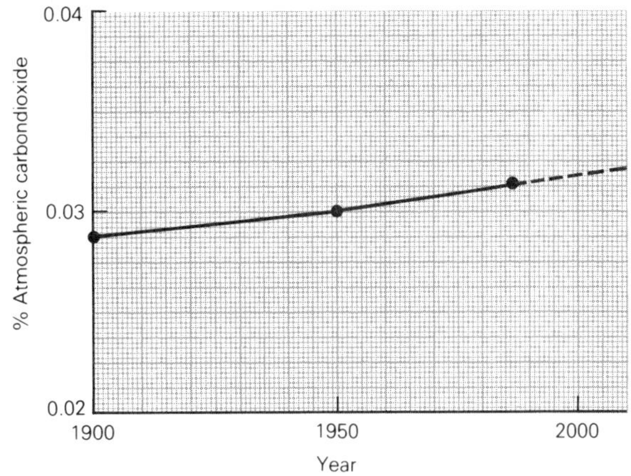

(ii) It may be useful to discuss the techniques required in writing for a specific audience before pupils tackle this. The main points of the account should be along the following lines:

'Scientists have noticed that the amount of carbon dioxide in the air is slowly going up. This is because we are burning more and more fuels like coal, petrol and wood. When these fuels burn carbon dioxide is made, and this gas goes into the air. Sunlight has to pass through the air to reach the ground, but it is not stopped by the carbon dioxide. The sunlight warms the ground, which will then give off heat energy. The carbon dioxide in the air takes in some of this heat. Instead of the heat going back into space from the Earth, some of it is trapped by this carbon dioxide blanket. The more carbon dioxide there is, the more heat will be kept back, so the warmer the Earth will become. This is called the "greenhouse effect", because the glass in a greenhouse lets light through but stops some of the heat escaping, making the inside much warmer than the outside.'

26 The nitrogen cycle in a meadow

Context

The nitrogen cycle is a relatively complex process and is generally only fully accessible to more able pupils. This question requires pupils to study a somewhat unfamiliar representation of the processes involved, thus helping to enhance their understanding.

Knowledge required

b The main stages of the nitrogen cycle.

d Fertilisers.

Skills

a, b Data extraction

c Data interpretation

d Application

Level

a Mod **b** Mod **c** Mod **d** Mod

Answers

a 3

b 3

c P_1 = photosynthesis; P_2 = feeding; P_3 = ammonification; P_4 = nitrification; P_5 = leaching; P_6 = denitrification; P_7 = nitrogen fixation; P_8 = lightning fixation.

d (i) Nitrogen fertiliser.
(ii) As plants are harvested, nitrogen compounds are taken away and therefore are not replaced by the nitrogen cycle.

27 The nitrogen cycle budget

Context

This question considers the ways in which nitrogen is gained and lost during the nitrogen cycle, and the effects of human influence. The data is presented in ways common in reports and texts, but possibly unfamiliar to pupils.

Knowledge required

d Processes in the nitrogen cycle.

Skills

a Data extraction

b Data interpretation

c Data handling (calculation) and Application

d Reasoning (prediction)

Level

a Mod **b** Mod **c** Diff **d** Diff

Answers

a 14 million tonnes.

b Because these represent loss of nitrogen from the ecosystem.

c (i) Net balance $= (14 + 30 + 4 + 30 + 14) - (43 + 30)$
$= 92 - 73$
$= 19$ million tonnes

(ii) Net balance $= (14 + 30 + 4) - (43 + 30)$
$= 48 - 73$
$= -25$ million tonnes

d Credit should be given for a sensibly argued answer, e.g. run-off would be reduced because the reduced application of fertilisers would mean there would be less nitrate to pass into rivers.

Denitrification might also be reduced if there were less nitrate available.

28 Nitrogen fixation in root nodules

Context

This question takes a closer look at the commercially important process of nitrogen fixation. It also provides an example of symbiosis.

Knowledge required

a to **h** require knowledge of basic processes in plants, i.e. photosynthesis, growth and translocation.

Skills

a, **b** Observation

c, **d** Data interpretation

e to **h** Reasoning (inference and hypotheses)

Level

a Easy **b** Easy **c** Easy **d** Easy+ **e** Mod **f** Mod **g** Mod **h** Mod

Answers

a 13

b (i) 5 (15 leaflets)
(ii) 10 (30 leaflets)

c B

d More leaves; longer stalks; more roots; 'better growth'.

e Any nitrogen compound, e.g. nitrate, ammonia or amino acids, would be a reasonable suggestion. Ammonia appears to be the main compound assimilated by the host cells.

f Sugar or other carbohydrate.

g It could be suggested that ions, such as nitrates, would be transported in the xylem. Alternatively, organic compounds would be translocated in the phloem.

h Products of photosynthesis would be translocated in the phloem from the leaves to the roots.

29 Fungal symbiosis

Context

This question provides another example of the use of radioactive isotopes as tracers, and requires pupils to interpret unfamiliar evidence.

Knowledge required

b, **c** Photosynthesis.

Skills

a, **b** Data interpretation

c Reasoning (explanation)

Level

a Mod **b** Diff **c** Diff

Answers

a When the mycorrhiza fungus was present, the number of grains per ear of maize increased more than ten-fold; the mass of each grain also increased by about eight-fold.

b After 12 hours the fungal threads contained radioactive sugars. These must have contained the ^{14}C from the $^{14}CO_2$, which suggests that the fungi were obtaining the sugars made by the photosynthesising pine seedling.

c The pine seedling absorbs $^{14}CO_2$ through the stomata. During photosynthesis this $^{14}CO_2$ is combined with water to make sugars. The sugars pass from the leaves to the roots in the phloem and are then absorbed by the fungal threads.

30 A lawn problem

Context

Mineral requirements of plants and fertilisers. The question puts this in the context of a common application.

Knowledge required

c Fertilisers as source of mineral ions.

d Osmotic effects of high ion concentration.

Skills

a, **b** Data extraction

c, **d** Reasoning (explanation)

Level

a Easy **b** Mod **c** Mod **d** Mod+

Answers

a Nitrogen, phosphorus, potassium.

b It contains the three elements in the ratio of 7:7:7 (or, strictly, the constituent compounds).

c So that there is a good supply of ions present when the seed germinates to ensure rapid early growth of the grass.

d Possible reasons include: high ion concentration: may be toxic; inhibit growth processes; upset osmotic balance; prevent or reduce uptake of water by osmosis.

31 The compost heap

Context

This question considers the procedures involved in the investigation of the organisms in a compost heap. The intention is that pupils should think out the difficulties in obtaining accurate data in this type of investigation.

Knowledge required

None.

Skills

a to **d** Experimental procedures

e Data handling (extraction and tabulation)

f Data handling (extraction and calculation)

g Data handling (graphs)

Level

a Mod **b** Mod **c** Mod **d** Diff **e** Mod
f Diff **g** Diff

Answers

a Light and warmth from the lamp stimulate the organisms to move down through the soil and fall into the meths.

b The animals have to be killed to prevent the carnivores eating the herbivores, and to preserve them.

c To make the investigation fair and to make the results from different groups comparable.

d (i) Advantage – a wide variety of organisms can be collected, even if they do not respond to light or warmth.
Disadvantages – small or well-camouflaged organisms may be missed; slow
(ii) Advantages – possible to collect organisms which are difficult to pick out by hand; less time-consuming.
Disadvantage – larger organisms cannot be collected in this way as they cannot pass through the gauze mesh.
(Clearly, any sensibly argued advantage or disadvantage should be credited.)

e

	Herbivores	Carnivores
Hand-sorting	171	51
Tullgren funnel	506	18

f 3 kg of compost searched by hand – divide by three for 1 kg; 50 g of compost in Tullgren funnel – multiply by twenty for 1 kg.

Herbivores		Carnivores	
Species	Total per kg of leaf litter	Species	Total per kg of leaf litter
Woodlice	5	Centipedes	3
Snails	4	Spiders	8
Millipedes	6	Harvestmen	2
Slugs	3	Beetle larvae	2
Springtails	5700	Ground beetles	2
Bristletails	2940		
Nematodes	14		
Leather-jackets	1		
Mites	1240		
Maggots	4		
TOTAL	9917	TOTAL	17

g Various types of diagrams are possible, e.g. a pie-chart or pictogram. A 'pyramid of numbers' is a likely solution, using a suitable area for a given number of individuals, e.g. 1 cm^2 = 100 individuals.

32 New House Farm – a managed ecosystem

Context

This problem examines the economics of operating a farm. It is based on data from an actual farm, and therefore offers a realistic insight. Pupils should appreciate that agricultural practices remove resources from the natural cycles and that these must therefore be replenished in some way.

Knowledge required

d Uses of fertilisers and chemical sprays (i.e. herbicides and pesticides).

e Potatoes are grown from tubers.

Skills

a Data extraction

b, **c** Data handling (calculation) and Application

d to **f** Applications (economic and technological)

Level

a Mod **b** Mod **c** Mod **d** Mod+ **e** Mod
f Diff

Answers

a

The total cost of running the farm	
Item	Total cost (£)
Rent	4200
Seed	6000
Fertilisers	9000
Chemical sprays	5000
Beef calves	3000
Wages	20000
Machinery	15000
Diesel	5000
b(i) TOTAL	67200

a

Money from sale of produce	
Item	Total cost (£)
Barley	24000
Wheat	13200
Oilseed rape	18000
Potatoes	5500
Beef cattle	11000
Pigs	800
Lamb	100
Turkeys	36
Eggs	120
b(ii) TOTAL	72756

c Profit = £5 556.

d Fertilisers are needed to replace minerals in the soil removed by the previous year's crop, and then to ensure good growth of the new crop.

Chemical sprays are needed to kill weeds which would compete with the crop for soil nutrients and light, and to kill parasites and insect pests which could damage the growing crops.

If this money were not spent the crops would be reduced and the returns would be less.

e When the farmer harvests potatoes he keeps back those too small to sell and plants them for the next season's crop.

f Successive crops of oilseed rape would tend to impoverish the soil by removing all the minerals, whereas rotation of crops helps to enable different minerals and different soil layers to be used. Growing only one crop would make it very susceptible to disease and pests, and also there would be financial problems if the price of the crop were to fall, for example because too many farmers turned over to oilseed rape so that supply exceeded demand.

UNIT 5

Man and the environment

33 Fertiliser pollution

Context

Human influences on ecosystems. The question refers to the process of eutrophication, resulting from excessive drainage of fertilisers into rivers and the sea.

Knowledge required

c Algae. Mineral salts needed for growth.

d Decay bacteria use oxygen.

Skills

a, **b** Data extraction

c Reasoning (prediction)

d Reasoning (explanation)

Level

a Easy **b** Easy **c** Easy **d** Diff

Answers

a 148 parts per million.

b 10% saturated.

c More rapid rate of growth, or increased population.

d The large population will eventually die, due either to depletion of other resources, e.g. access to light, space etc., or to natural causes. As the algae decompose there will be an explosion in the bacterial population, which will deplete the oxygen concentration of the water.

34 Pollution indicators

Context

Pollution by sulphur dioxide.

Knowledge required

None.

Skills

a Data interpretation

b Reasoning (inference)

c Experimental design and Data interpretation

Level

a Easy **b** Mod **c** Mod+

Answers

a The nearer the city centre, the fewer the number of lichen species.

b As the concentration of sulphur dioxide rises the number of lichen species declines, suggesting that many species are killed off or fail to grow in the higher concentrations of sulphur dioxide. Other reasons may be suggested, but these would not be supported by the data given.

c Pupils should appreciate that the number of species present is proportional to the concentration of sulphur dioxide. Therefore they would need to carry out a survey of species present. They could then correlate this number with the appropriate concentration of sulphur dioxide shown in the graph.

35 Whale hunting

Context

Populations. The data illustrates the effect of human predation and raises the issue of the need for conservation of whales, which pupils may wish to discuss.

Knowledge required

None.

Skills

a Data handling (graphs)

b Data interpretation

c Reasoning (inference)

Level

a Easy **b** Easy **c** Easy+

Answers

a Graph. Pupils need to be careful to use a linear scale for time. Since population numbers in intervening years are not known, straight lines should be used to connect the points.

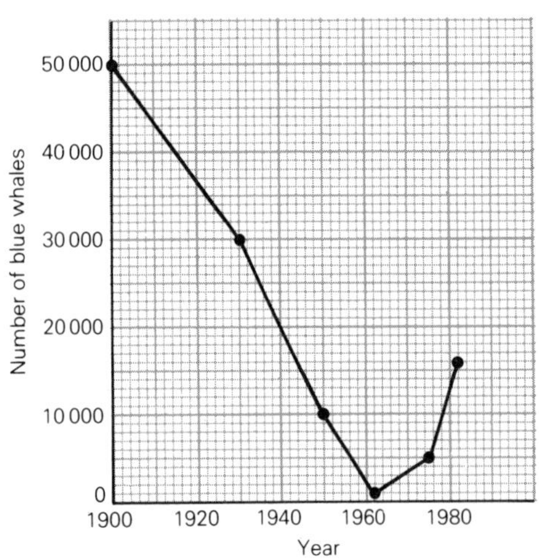

b The population fell steadily until about 1960. Since then there has been a recovery in numbers.

c The fall in the first half of the century was the result of whale-hunting. Since 1966 the blue whale has been officially protected, so the recovery in numbers can be assumed to be the result of reduced 'human predation'.

36 Culling

Context

The question focuses on the idea of natural populations being adapted to their customary environment, and the disadvantages of using artificially selected breeds of domestic animals.

Knowledge required

Understanding of adaptation and selection.

Skills

Application.

Level

Mod+

Answer

As a result of natural selection a natural population is likely to be better able to make use of the indigenous food plants, withstand environmental conditions and resist disease.

37 Air pollution

Context

Pollution.

Knowledge required

c and **d** Sources and effects of major pollutants. Pupils might be encouraged to look these up.

Skills

a Data handling (calculation)

b Data handling (graphs)

c Knowledge

d Application

Level

a Easy **b** Mod **c** Mod **d** Mod+

Answers

a

Gas Pollutant	Percentage	Angle for pie-chart
Carbon monoxide	55.5	200°
Sulphur oxides	12.8	46°
Nitrogen oxides	8.7	31°
Hydrocarbons	13.2	48°
Particles, e.g. carbon	9.8	35°

b Pie-chart. Pupils may convert the order of sectors into rank order of magnitude.

c Carbon monoxide – vehicle exhausts
Sulphur oxides – combustion of coal and oil, e.g. in power stations
Nitrogen oxides – vehicle exhausts
Hydrocarbons – burning coal, and vehicle exhausts
Particles – coal fires and coal-fired power stations; 'soot'
(They may all be grouped together as pollutants resulting from combustion of fossil fuels.)

d Carbon monoxide – toxic effects, by combining with haemoglobin
Sulphur oxides – acid rain, and hence its effects on trees, freshwater life, building stone, etc.
Nitrogen oxides – 'smog', lung damage, etc.
Hydrocarbons – liver damage and possibly cancer
Particles – fog, blackening of paintwork and stone, etc.; lung damage.

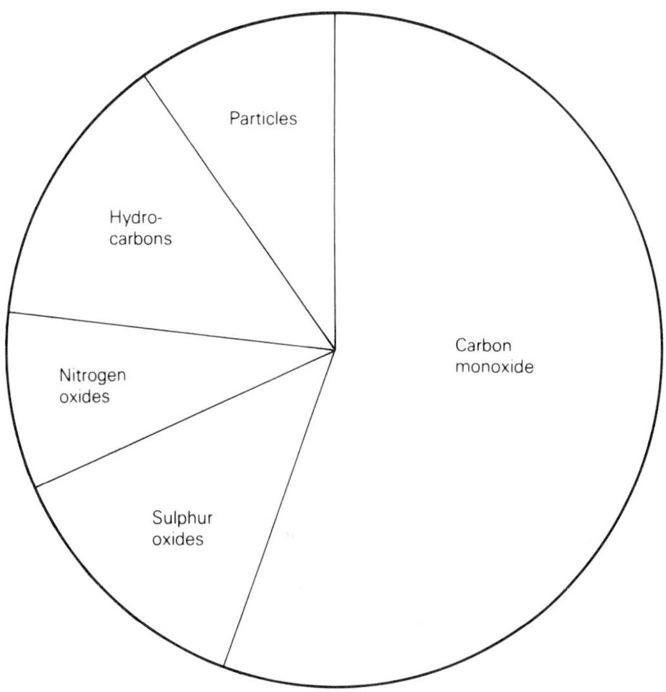

38 Lead poisoning

Context

Pollution. Lead affords a relevant and important example of the toxicity of heavy metal ions. The change-over to the use of unleaded petrol is likely to take several years.

Knowledge required

e Lead in petrol.

Skills

a to **c** Data extraction and handling (calculation)

d Reasoning (hypotheses)

e Knowledge (environmental/social)

Level

a Mod **b** Mod **c** Mod **d** Mod+ **e** Easy

Answers

a 38 mg

b 18.4 mg

c Food – 5.2%
Water – 10.0%
Air – 40.0%
Smoke – 40.0%

d The most obvious suggestion is that the lung membrane, through which lead is absorbed from air and smoke, is more permeable or thinner than the gut membrane. Other possible reasons would be connected with the nature of the lead compounds in the different sources; active transport mechanisms; or that much of the lead in food is in indigestible components which are eliminated.

e Car exhausts, i.e. leaded petrol.

39 DDT

Context

Pollution by pesticides. Although DDT is now banned in Britain it is still widely used in many Third World countries and residues are still detectable throughout the world. The principles illustrated are also applicable to other pesticides.

Knowledge required

Food chains and trophic levels.

Skills

a Reasoning (explanation)

b, **c** Data handling (calculation)

d Data interpretation

e Reasoning (inference)

f Experimental design

g Application (economic and technological)

Level

a Mod **b** Mod **c** Mod **d** Mod **e** Mod
f Mod+ **g** Mod+

Answers

a Each individual consumer eats many individuals of its food organism. The DDT is absorbed into the body of the consumer and stays there; therefore the concentration rises.

b Small fish × 5.75
Large fish × 9
Birds of prey × 12.75

c × 800 (the overall concentration factor for birds of prey is therefore × 528 000).

d Widespread use of DDT began in the mid-1940s. The graphs show that from this time the thickness of the eggshells of both peregrine falcons and sparrowhawks decreased significantly and rapidly.

e The concentration of DDT would be higher in birds of prey and therefore the effect would be likely to be greater.

f The most likely suggestion is to take two captive samples of the birds; feed one group on DDT free food and the other with food containing amounts of DDT equivalent to that found naturally. The thickness of the shells could then be compared. (Such investigations have in fact been carried out with, for example, kestrels, and have confirmed the effect of DDT on shell thickness.)

g Such factors as toxicity, both to the insects and other organisms including human beings; selectivity; stability; absorption by insects; ease of application; cost; effects on common materials, e.g. cloth; etc.

40 Biological control

Context

This question provides pupils with information about a method of biological control commonly used by commercial greenhouse owners and gardeners.

Knowledge required

a Understanding of the term 'life cycle'.

b Sap, as sugary contents of phloem. Photosynthesis.

Skills

a Data extraction and handling (diagrams)

b Reasoning (explanation)

c Data interpretation

d Data extraction and handling (calculation)

e Data interpretation

f to **j** Application (economic and technological)

Level

a Mod **b** Mod **c** Mod **d** Mod **e** Mod
f Mod+ **g** Diff **h** Mod **i** Diff **j** Diff

Answers

a (i)

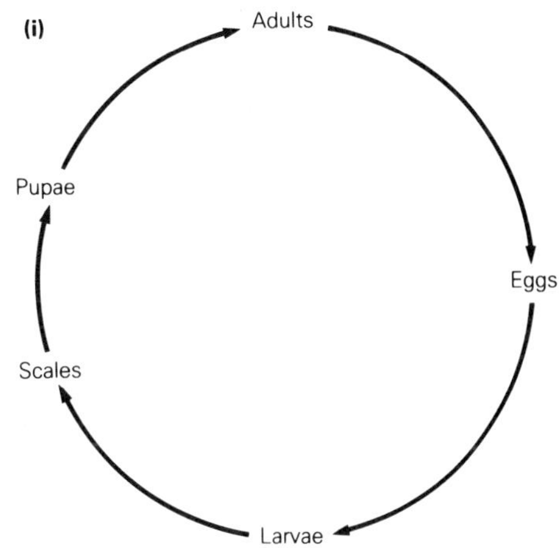

(ii)

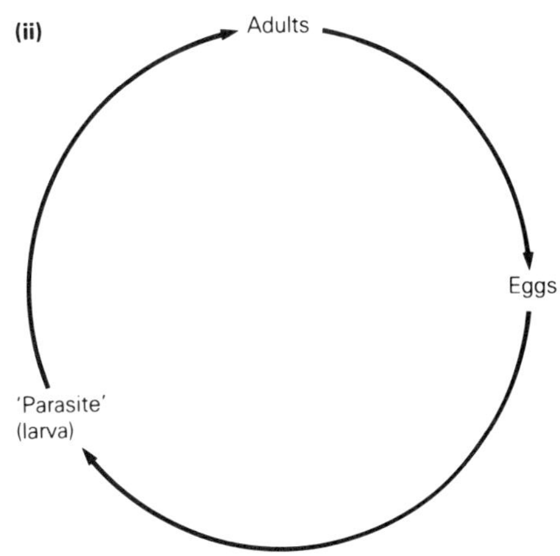

These 'life cycles' are as they can be deduced from the information given. *Encarsia* does include a larval and pupal stage inside the scale.

b (i) The whiteflies remove sugars, and other organic compounds, from the phloem. They therefore deprive the plant of some of its photosynthetic product and thus reduce its rate of growth.

(ii) The reduction in light entering the leaves will reduce the rate of photosynthesis, and then further reduce the amounts of sugar available.

c The scales, when black, are known to contain larvae. They are also nearer to hatching than eggs and therefore to being ready to attack the whitefly. If adults were sold they might already have laid eggs or be too old before being set on the whitefly. Scales are also easier for gardeners to handle.

d $800 \times 50 = 40\,000$.

e Permethrin kills the highest percentages of both adults and pupae and is cheap.

f Size of greenhouse; number and 'area' of plants; density of whitefly infestation.

g None of the insecticides kills all the whitefly, and especially not the younger stages. As more adults emerge treatment has to be repeated.

h The pupae do not feed and are less likely to take the insecticide into the body. They may also have a less absorptive surface layer, since the pupal case will be shed during hatching.

i Once placed in the greenhouse the *Encarsia* should need no further attention and should keep reproducing until the infestation is controlled. There would be no need for regular spraying and therefore labour costs could be reduced. There would also be no risk of contamination of the crop by the insecticides.

j Insecticide sprays would have to be used, as the *Encarsia* could not be controlled outside.

UNIT 6

Structure and chemistry

41 Cells

Context

Differences between plant and animal cells.

Knowledge required

a, **b** Familiarity with cell structure.

e Photosynthesis requires carbon dioxide.

Skills

a, **b** Observation

c Measurement

d Observation and data interpretation

e, **f** Reasoning (inference)

g, **h** Observation and Data handling (diagrams)

Level

a Easy+ **b** Easy+ **c** Mod **d** Mod
e Diff **f** Diff **g** Easy+ **h** Easy+

Answers

a PALISADE CELLS	WINDPIPE CELLS
cell wall	no cell wall
chloroplasts	no chloroplasts
nucleus not visible	central nucleus visible
no hairs/cilia	hairs/cilia on one edge
large (over 100 μm long)	smaller (about 50 μm long)

b Cheek cells have no chloroplasts, are smaller and more flattened.

c (i) 44.2 μm (ii) 17.6 μm (iii) 80.1 μm (iv) 39.6 μm (v) 5.2 μm

d The palisade cell is long and thin, and roughly circular in cross-section.

e To allow more rapid diffusion of carbon dioxide into the cells for photosynthesis.

f To reduce water loss, and to form a protective layer, preventing bacteria, etc., entering the leaf.

g, h Drawings as produced by pupils

42 Water

Context

The importance of water as a constituent of living organisms, and the properties which make it particularly suitable.

Knowledge required

d Water relations of a mammal.

f Decay.

Skills

a Data extraction

b Data handling (calculation)

c Data interpretation

d Knowledge and Application

e Reasoning (explanation)

f Data handling (calculation) and Reasoning (explanation)

g Application

h Experimental design and Application (economic)

Level

a Easy **b** Easy **c** Mod **d** Easy **e** Easy
f Mod **g** Mod+ **h** Mod+

Answers

a (i) Brain (ii) Fat

b (i) 49 kg (ii) 22.5 kg (iii) 4 kg
(iv) 1.19 kg

c Less, because fat has a very low percentage of water. Therefore, the total water content of the body is only increased by a small amount compared with the total increase in mass. This may be clearer to pupils if a simple calculation is done,
i.e. water content = $(70 \times 70\%) + (20 \times 20\%)$
$= 53$ kg
Therefore proportion $= \frac{53}{90} \times 100\% = 58.9\%$

d (i) Sweat, urine, faeces, breath (and small amounts in tears, etc.).
(ii) Drinking, and substantial amounts in food.
(iii) Hot conditions increasing sweating; exercise raising rate of breathing and body temperature and hence sweating; high intake of water; diarrhoea; or any other sensible suggestion.

e Chicken has a high water content, much of which will evaporate during roasting.

f (i) 50 g
(ii) Bacteria and other organisms cause meat to go bad; these require water to survive.

g (i) Substances can only react in solution, and need to be dissolved for transport.
(ii) Light can penetrate cells for photosynthesis, or water to allow aquatic organisms to photosynthesise. Transparency is also important in light-sensitive organs such as eyes.
(iii) Organisms can maintain a relatively stable temperature as external temperatures fluctuate. Ponds, lakes, the sea, etc., are also relatively slow to heat up and cool down.
(iv) Water is relatively slow to evaporate, hence helping organisms to conserve water; body temperature can be lowered by evaporation of water.
(v) Ice forms on *top* of ponds, lakes, etc., so organisms can continue to live in the liquid water below.

h (i) Take two batches of apples; weigh each batch or apple; spray one batch with the substance; reweigh at fixed intervals. Pupils may also suggest comparisons based on appearance, but should be encouraged to select clearly measurable quantities.
(ii) Toxicity to animals and human beings, including effects on sprayers who handle large quantities; possible side effects, e.g. allergies, damage to eyes, etc.; taste and palatability of apples; cost of spray; capital costs of spraying equipment; labour costs of spraying; possible deterioration due to handling or during transport; extra value of keeping apples before selling.

43 What are we made of?

Context

The major biochemical compounds constituting living organisms, and the uses of minerals.

Knowledge required

This question requires some background knowledge of the chemistry of elements and compounds.

b to **d** Elemental composition of carbohydrates, fats, proteins and water.

e Atomic masses of hydrogen and oxygen.

f, **g** Functions of calcium and iron in human beings.

Skills

a Data handling (graphs)

b to **e** Knowledge and Data interpretation

f, **g** Knowledge

Level

a Mod **b** Mod **c** Mod **d** Mod **e** Diff
f Easy **g** Easy

Answers

a

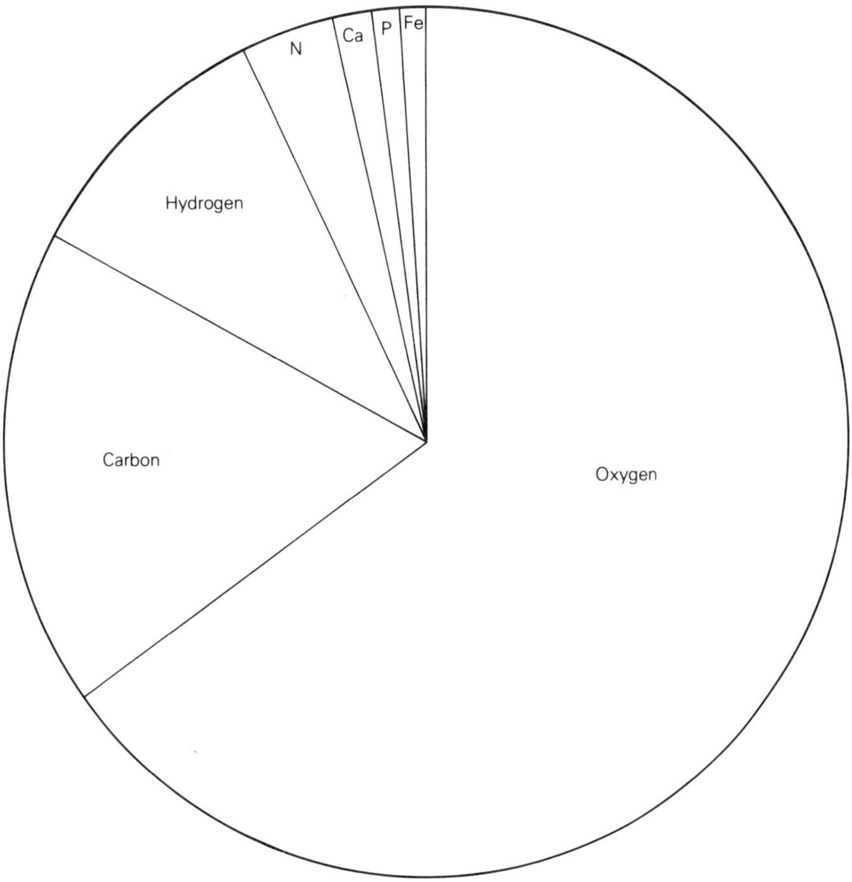

b Fats and proteins.

c Proteins.

d Water.

e The atomic mass of oxygen is 16, whereas that for hydrogen is 1. Thus in water the mass of oxygen is eight times as great as that of hydrogen.

f Bone.

g Blood (red blood cells).

44 Food tests

Context

Simple biochemical tests.

Knowledge required

Iodine, Benedict's and biuret tests.

Skills

a to **f** Data interpretation

Level

a Easy **b** Easy **c** Easy **d** Easy **e** Easy
f Easy

Answers

a

Food	Substances contained		
	Starch	Sugar	Protein
A	√	√	×
B	×	√	√
C	×	√	×
D	√	×	√
E	√	√	√

b A, D and E

c C

d A and E

e E

f D

45 Effects of temperature on an enzyme

Context

Principles of enzyme action and digestion. The question concentrates on the experimental procedures involved in the familiar investigation of the effect of temperature on the activity of amylase. It can be used to reinforce understanding of the method and of the problems which arise in such investigations.

Knowledge required

It is most appropriate if pupils are already familiar with the basic experimental procedures in testing for rate of starch digestion.

c, **d** Effect of high and low temperatures on enzymes.

e Human body temperature as 37 °C.

Skills

a Data handling (graphs)

b Data extraction

c to **e** Data interpretation

f Experimental procedure

g Experimental design

h Experimental procedure

Level

a Easy **b** Easy **c** Mod **d** Mod **e** Mod
f Mod **g** Mod+ **h** Mod

Answers

a Graphs. Pupils may either join points by straight lines, or draw a curve of best fit. The anomalous point should not be included.

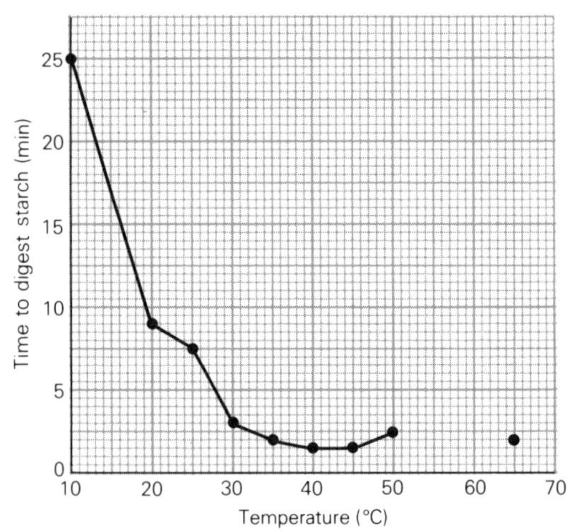

b Approximately 40 to 45 °C, according to pupil's graph.

c Low temperature slows down rate of chemical reaction (lower kinetic energy of molecules).

d The enzyme has been denatured at this temperature so is not able to catalyse the breakdown of starch.

e Rate of digestion would increase slightly.

f (i) K
(ii) The most likely reason is that the enzyme had not been fully denatured when the enzyme and starch were mixed. Pupils may also suggest variability in the enzyme.

g (i) Measure a fixed volume (e.g. 10 cm^3) of starch solution into a test-tube; measure a smaller volume of enzyme (e.g. 2 cm^3) into a second tube; put both tubes into a water-bath at 50 °C for several minutes until both solutions reach 50 °C; mix enzyme with starch solution; every half minute remove a drop and place one drop of iodine solution on a tile.
(ii) Apparatus chosen will vary according to precise procedure adopted. A suitable list would include: measuring cylinder or syringe; two test-tubes; beaker, burner, mat, tripod, gauze, thermometer (or water-bath), dropper or glass rod, white tile, clock.

h Add small amounts of crushed ice to cold water.

46 New and old spuds

Context

Principles of enzyme action. This question refers to a starch-digesting enzyme other than the ones found in the mammalian digestive system, and also illustrates an alternative method of assaying starch digestion.

Knowledge required

b The iodine solution test for starch. Principles of starch digestion.

c Breakdown of starch to sugar. Test for sugar.

e Role of potatoes as perennating organs.

Skills

a Experimental procedures

b Data interpretation

c Data interpretation and Experimental procedure

d Data interpretation

e Reasoning (explanation)

Level

a Easy+ **b** Mod **c** Mod **d** Mod+ **e** Diff

Answers

a Use pieces of potato of the same dimensions; take pieces from similar parts of the potato; keep both dishes in the same conditions, e.g. temperature, light; make sure that the starch concentration is the same in both dishes.

b There was no starch in the clear areas. Therefore the starch must have been digested by enzymes from the potato.

c (i) Sugar
(ii) e.g. Benedict's test – put sample of agar in test-tube; add Benedict's solution; boil gently; orange/red precipitate shows sugar present.

d No, since the pieces of potato were of different sizes.

e Old potatoes which are sprouting break down the starch reserves to sugars for translocation to growing points. In new potatoes the starch reserves are required for long-term storage.

47 Cheese-making

Context

The question relates to a biotechnological use of an enzyme, i.e. the use of rennin in cheese-making.

Knowledge required

None.

Skills

a Data handling (graphs)

b Data extraction

c Data interpretation

d Application (economic)

Level

a Easy **b** Easy **c** Easy **d** Mod

Answers

a Graph. Weaker pupils may need guidance on axes and plotting.

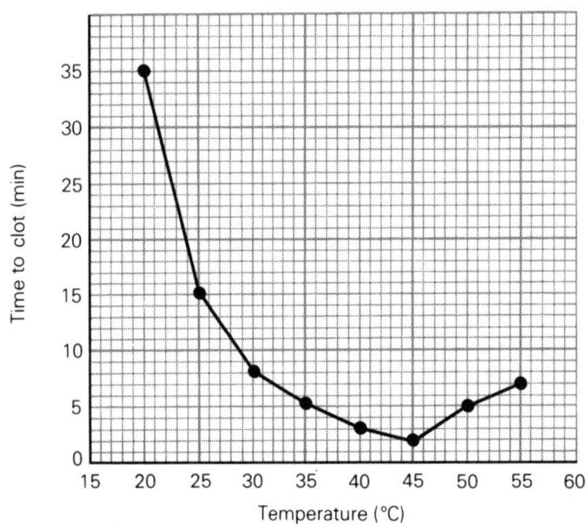

b Approximately 45 °C.

c Approximately 45 °C.

d The energy costs of maintaining the temperature. It might be cheaper to allow curd to be made at a slightly lower rate, but at a temperature which costs less to maintain. Pupils might also mention possible effects on taste, etc.

48 Brown potatoes

Context

Principles of enzyme action. This question illustrates the action of an enzyme which is not a digestive enzyme.

Knowledge required

b, c Enzymes, and factors affecting their activity.

Skills

a, b Data interpretation

c Reasoning (prediction)

d Experimental procedure

e Data interpretation

f Reasoning (hypotheses) and Experimental design

g Experimental design

h to **j** Reasoning (inference) and Application

Level

a Mod+ **b** Mod+ **c** Mod **d** Mod
e Mod+ **f** Diff **g** Mod+ **h** Mod **i** Mod
j Diff

Answers

a Potato juice contains something which makes catechol go brown.

b From tube C, boiling prevents the activity of the potato juice. Since boiling denatures enzymes, it is possible that an enzyme is involved.

c Increased rate of browning in tubes A and D, since temperature increases rate of reaction.

d B acts as a control to confirm that it is the potato juice that makes the catechol turn brown. D acts as a control to check that the potato juice is not the source of the browning effect. C may be considered as a control in the investigation of temperature dependence.

e Potato juice contains either small amounts of catechol or something else, which is turned brown by the enzyme in the juice.

f (i) The process requires air and therefore probably oxygen.

(ii) The experiment should test the hypothesis suggested in part (i). In principle it should involve mixing catechol and potato juice in the absence of, or with reduced amounts of, air or oxygen, e.g. by covering the mixture with oil, bubbling nitrogen through, or comparing an aerated and unaerated tube.

g Prepare an extract of banana juice; place a fixed volume, e.g. 5 cm^3, of catechol in a tube; add a similar volume of banana juice; compare rate of browning with control.

h The boiling will destroy the enzyme which causes browning.

i To prevent further softening or cooking while still hot, or other suitable suggestion.

j Peas are much smaller than new potatoes. Therefore the enzymes in the middle of the pea will be raised to a high temperature and denatured much more quickly.

UNIT 7 Plant nutrition

49 Leaves

Context

Photosynthesis and leaf structure. The first parts of the question relate to the standard starch test as an indicator of photosynthetic activity in the light. The remainder consider the structure and adaptations of the typical leaf in more detail.

Knowledge required

a, **b** Procedures for carrying out starch tests on a leaf.

d, **e** Chloroplasts.

g Requirements for photosynthesis.

Skills

a, **b** Experimental procedure

c Data interpretation and handling (diagrams)

d Observation and Data handling (tabulation)

e Reasoning (hypothesis)

f Measurement

g Reasoning (explanation)

Level

a Easy **b** Easy **c** Mod **d** Easy+ **e** Mod **f** Mod **g** Easy+

Answers

a (i) Leave the plant in darkness for at least 24 hours.

(ii) To show that any starch present on testing had been produced in the light, and was not already in the leaf.

b Dip the leaf in boiling water, and then place in boiling ethanol (using a water-bath for safety) until colourless.

c

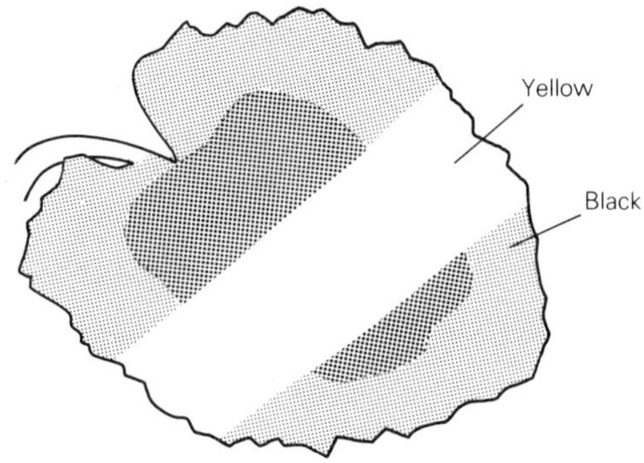

d

		Palisade layer	Spongy layer
Number of chloroplasts	Cell 1	16	6
	Cell 2	17	8
	Cell 3	18	4
	Cell 4	14	7
	Cell 5	15	7
	Total	80	32
	Average	16.0	6.4

e Palisade layer receives more light, or other sensible suggestion.

f (i) Approximately 275 micrometres.
(ii)

LAYER	THICKNESS (μm)
Upper epidermis (excluding cuticle)	20.25
Palisade layer	80.85
Spongy layer	130.14
Lower epidermis	20.00

g Light has less far to penetrate; carbon dioxide has less far to diffuse; large surface can be exposed to light without having to support a large mass.

50 Temperature and photosynthesis

Context

Factors affecting the rate of photosynthesis. The question also asks pupils to think about the economics of increasing the rate of photosynthesis in a greenhouse.

Knowledge required

d, **e** Factors affecting the rate of photosynthesis. For some pupils it may be necessary to explain the term 'arbitrary units', used here to avoid the complication of 'mm^3 CO_2 per cm^2 of leaf per hour'.

Skills

a, **b** Data extraction

c Data interpretation

d Application (economic)

e Application

Level

a Easy **b** Easy **c** Easy+ **d** Mod+
e Mod

Answers

a 23 units

b 22 °C to 29 °C

c As temperature increases up to about 21 °C the rate of photosynthesis increases proportionately. Above this temperature the rate remains constant up to about 29 °C and at higher temperatures the rate declines. Pupils may be able to link the proportional rise to rate of chemical reaction; the constant rate to a limiting factor, such as light or carbon dioxide concentration; and the fall to the effect of heat on the enzymes.

d From the graph there would be no advantage in raising the temperature above about 22 °C as there is no increase in rate above this.

e Additional light, as winter days are short and often very dull. It might also be advantageous to supplement the carbon dioxide concentration.

51 Stomata

Context

Photosynthesis and leaf structure. The data contradict some of the simplified accounts of stomatal opening and closure given in the textbooks.

Knowledge required

Stomata, as leaf pores.

Skills

a to **c** Observation and data extraction

d Data interpretation

Level

a Easy **b** Easy **c** Easy **d** Diff

Answers

a (i) Between 6 and 7 a.m.
(ii) Between 7 and 8 p.m.
(iii) Between 11 p.m. and midnight
(iv) Between 8 and 9 p.m.

b (i) Potato.
(ii) About 8 hours.

c 11 hours.

d The hypothesis is supported by the maize, but the potato stomata are open for much of the night as well and therefore contradict the hypothesis. The hypothesis is not true for all species.

52 How do stomata open and close?

Context

The question concentrates on measurement and scaling skills. Reference is made to the changes in guard cell shape resulting from osmosis. With able pupils teachers may wish to consider possible causes of the osmotic effects, but care should be taken not to instil over-simplified ideas. The terminology used in relation to osmosis is confusing; pressure potential is now recommended as an alternative to the term 'turgor pressure'.

Knowledge required

Stomata as pores in leaves.

c, **d** Principles of osmosis.

Skills

a, **b** Measurement and Data handling (calculation)

c Data interpretation

d Reasoning (explanation)

Level

a Mod **b** Mod **c** Mod **d** Diff

Answers

a 3 μm

b

	MEASUREMENT (μm)	
	Stoma closed	Stoma open
Depth of guard cell	7.5	12.9
Width of guard cell	13.3	12.1
Depth of vacuole	5.8	7.9
Width of vacuole	7.5	7.1

c (i) Not turgid – there is little water in the vacuole.
(ii) Turgid.

d The fact that when the guard cells are turgid the stoma is open suggests that, as water enters, the cells expand and leave a gap between them. The cells are taller but narrower when turgid. As can be seen from the diagram, when the cells lose water they collapse because of the thin walls bordering the stoma, causing the pore to be closed.

53 Oxygen production in photosynthesis

Context

Photosynthesis. The data require pupils to interpret an unfamiliar piece of evidence relating to oxygen production.

Knowledge required

Photosynthesis.

Skills

a Observation

b, **c** Data interpretation

d Experimental design and Application

Level

a Easy **b** Easy+ **c** Easy+ **d** Diff

Answers

a (i) Most bacteria are close to the chloroplast.
(ii) The bacteria are clustered round the spot of light shining on the chloroplast.

b The chloroplast. The bacteria move to high oxygen concentrations. They cluster round the chloroplast when it is illuminated, suggesting that this is where oxygen concentration is highest, and hence that the oxygen is produced by the chloroplast and not other parts of the cell.

c In darkness the bacteria do not cluster round the chloroplasts. In light they do, even if only a small part is illuminated. Therefore oxygen production must occur only in light.

d (i) Illuminate *Spirogyra* with spots of light, using red, yellow, green or blue filters over the light.
(ii) Count the numbers of bacteria clustering in each coloured spot.
(iii) If one colour has more bacteria it would show more oxygen is being produced and therefore more photosynthesis is taking place. It might be advantageous to illuminate plants with the colour which gives the best result. Comparison would, however, have to be made with white light.

54 Light intensity and photosynthesis

Context

Relationship between photosynthesis and respiration, and the compensation point. Pupils are required to study carefully the fairly complex instructions involved in carrying out the investigation. The results of the investigation illustrate a physiological adaptation of plants to their environment.

Knowledge required

Processes of photosynthesis and respiration.

Skills

a to **f** Experimental procedures

g Data interpretation

h Data handling (graphs)

i, **j** Data interpretation

Level

a Diff **b** Diff **c** Diff **d** Diff **e** Diff **f** Diff
g Diff **h** Diff **i** Diff **j** Diff

Answers

a To make sure there were no air bubbles in the apparatus.

b Because of the very small diameter, a small change in volume produces a large movement in the tube, which increases the accuracy of measurement.

c To allow for movement in either direction.

d The same apparatus set up without the leaf.

e To prevent heat from the lamp raising the temperature of the water in the syringe.

f To make the results comparable, since individual leaves vary in size.

g (i) The water is being forced along the tube because the rate of photosynthesis is greater than the rate of respiration and oxygen is being released.
(ii) The rate of photosynthesis is less than the rate of respiration, and there is a net loss of oxygen from the leaf air spaces into the leaf cells. Therefore the volume of gas inside the leaf decreases and the water is drawn into the air spaces.

h Graph.

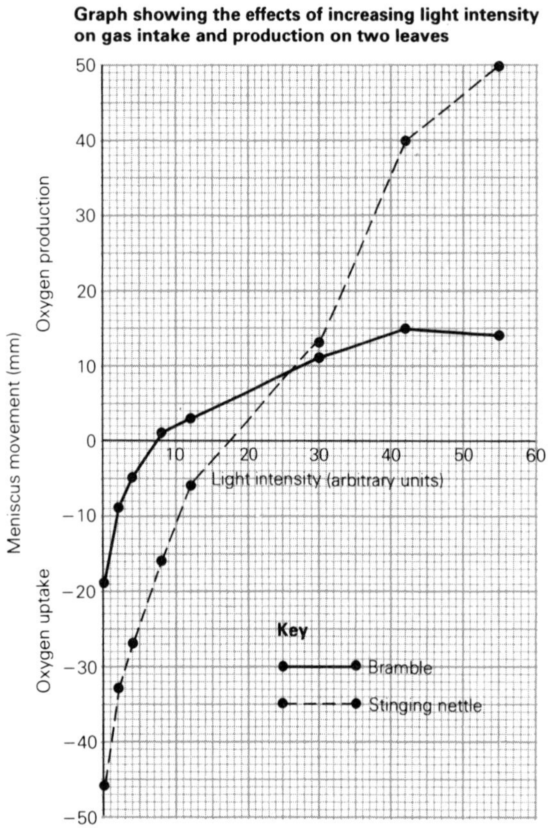

i Bramble: 7.5 arbitrary units.
Stinging nettle: 17.5 arbitrary units.

j Bramble. The rate of photosynthesis is greater than the rate of respiration at lower light intensities than in the stinging nettle. Therefore it will be able to make a net gain in food at lower light intensities, i.e. in relatively shady conditions.

55 Can we improve on nature?

Context

The question deals with the interaction between the factors which affect rate of photosynthesis, and hence covers the idea of one factor limiting the rate. This leads on to the technique of increasing carbon dioxide concentration in order to improve yields of greenhouse crops.

Knowledge required

Photosynthesis as a process of plant nutrition.

Skills

a, **b** Data extraction

c to **e** Data interpretation

f Application (economic)

Level

a Easy **b** Easy **c** Mod **d** Mod **e** Easy
f Mod+

Answers

a H, i.e. 30 °C, 6 a.u. light, 0.13% carbon dioxide.

b Rate of photosynthesis increases in each case.

c Increasing the concentration of carbon dioxide can increase the rate of photosynthesis, and hence the rate of growth and the yield.

d The carbon dioxide concentration is the same in all four; increasing the light intensity or the temperature can increase the rate of photosynthesis, so carbon dioxide concentration must be limiting its rate.

e In both crops the yield is increased.

f Any sensibly reasoned answer should be acceptable, but some are more sensible than others. Economic considerations should form the basis of the judgement. Using paraffin heaters has the economic advantage of providing heat as well as carbon dioxide.

56 The 'Green Revolution'

Context

This question might be used in the context of solving the problems of world food provision, or of artificial selection as a means of improving agricultural yield.

Knowledge required

None.

Skills

a to **c** Data extraction

d Application (economic)

Level

a Easy **b** Easy **c** Easy **d** Mod+

Answers

a Taiwan – 2380 kg per hectare.
West Pakistan – 1530 kg per hectare.
Ceylon (Sri Lanka) – 1100 kg per hectare.

b Taiwan – 3040 kg per hectare.
West Pakistan – 1670 kg per hectare.
Ceylon (Sri Lanka) – 1300 kg per hectare.

c About 1964, or slightly before.

d The capital cost of providing the extra fertiliser required; the need for careful irrigation and flood control to prevent prolonged submersion, and possibly, therefore, the cost of work to provide this.

57 Food crops

Context

Plants as a source of food. The question illustrates the fact that different crop plants can provide totally different proportions of carbohydrate, fat and protein, and deals with micro-organisms as a potential source of protein for human consumption.

Knowledge required

Carbohydrates, fats and proteins as components of human diet.

f Energy losses at different trophic levels in food chains.

Level

a Easy **b** Mod **c** Easy **d** Mod **e** Mod+ **f** Mod **g** Mod+

Skills

a to **c** Data extraction

d Data interpretation

e Application (social)

f Reasoning (explanation)

g Application (social and economic)

Answers

a Banana.

b From the data, banana has the highest proportion of other constituents including fibre, and probably therefore has most fibre. Pupils should note, however, that on this evidence alone it is possible that fibre is a low proportion of the other constituents and that other crops might still have a higher content.

c

Rank order	Carbohydrate	Fat	Protein
1	rice	soya	blue-green alga
2	wheat	blue-green alga	soya
3	blue-green alga	wheat	wheat
4	banana	rice	rice
5	soya	banana	banana

d It is richest in protein, which is expensive and often in short supply in human diets. Since it grows naturally in ponds, it is easy to grow and harvest. It contains a high proportion of useful nutrients.

e (i) Acceptability, e.g. in terms of taste, appearance and texture.

(ii) Artificial fabrication into forms more like familiar foods, such as meat fibres; adding artificial flavours and colouring to improve palatability; mixing with other foods to disguise appearance and taste of raw SCP.

f In food chains energy is lost at each trophic level and only a small fraction of food is converted into new growth. By feeding SCP to animals an extra stage of energy loss is introduced, making the process more inefficient.

$$\text{SCP} \xrightarrow{\textit{loss}} \text{animal} \xrightarrow{\textit{loss}} \text{human beings}$$

g Answers to this may be developed at various levels. Increasing world population means ever-increasing demands for food and relatively little land available for expansion; climatic changes are also reducing available agricultural land in many parts of the world. Providing meat protein is inefficient for the reasons given in part **f**. For many people meat is too expensive as a source of essential protein. Objections on principle to eating meat may also be considered.

UNIT 8 Animal nutrition

58 Energy from sweets

Context

Energy value of food and uses of energy in the body. Although the calorie is not a unit of energy generally used in GCSE or other science syllabuses, it does have widespread currency in popular publications, and it is worthwhile to relate this unit to the joule and work done in science. The question illustrates how much care needs to be taken in using data from commercial magazines. It also provides practice in calculating the energy requirements of different activities. It is assumed that pupils will use calculators.

Knowledge required

General understanding of energy and its uses in living organisms.

Skills

a to **e**, **g** Data handling (calculation)

f, **h** Reasoning (explanation)

Level

a Easy **b** Mod **c** Mod **d** Mod **e** Mod
f Mod **g** Mod **h** Mod

Answers

a Next column.

b Revels; Ripple; Treets; Galaxy; Maltesers, Marathon, Twix; Bounty; Milky Way; Topic; Mars; Minstrels; Tunes; Opals; Lockets.

c (i) Lockets. (ii) Revels.

d 1179.4 kJ.

e 197 minutes = 3 h 17 min.

	Energy (kJ)	Energy per g (kJ)
Bounty	491.4	19.7
Galaxy	1524.6	21.8
Lockets	634.2	14.7
Maltesers	504	20.2
Marathon	504	20.2
Mars	1134	18.1
Milky Way	487.2	19.5
Minstrels	869.4	17.7
Opals	760.2	14.9
Revels	739.2	29.6
Ripple	646.8	25.9
Topic	974.4	18.4
Treets	1041.6	24.8
Tunes	571.2	15.4
Twix	504	20.2

f Maintaining body temperature; breathing movements; heartbeat; peristalsis; any other muscular movements, e.g. maintaining body posture, blinking; chemical processes, e.g. in growth; active transport, etc.

g 39 minutes.

h Much more muscular activity.

59 Milk

Context

Constituents of the human diet. The data illustrate some marked differences between human and cow's milk which could form a basis for discussion of the relative merits of breast and bottle feeding.

Knowledge required

None.

Skills

a to **e** Data extraction

f to **i** Data handling (calculation)

j Application

Level

a Easy **b** Easy **c** Easy **d** Easy **e** Easy
f Mod **g** Mod **h** Mod **i** Mod **j** Mod+

Answers

a Proteins, calcium.

b Lactose.

c 13 g

d 33 g

e 0.25 mg

f 833 g

g (i) Yes (15.6 g). (ii) No (408 mg).

h 1471 g (i.e. enough to provide 500 mg of calcium).

i 7143 g (i.e. over 7 litres!).

j In the first 6 months, from stores of iron already present in the infant's body. Later in other dietary supplements, e.g. from appropriate fruit and vegetables.

60 Square meals

Context

Balanced diet and human energy requirements. The question provides practice in handling a relatively large amount of data without the time-consuming activity of accumulating similar data for a personal diet, although exercises might be done to generate personal information for at least one meal.

Knowledge required

d, e Energy requirements and uses.

f Uses of calcium.

g Uses of iron.

Skills

a Data extraction

b Data handling (calculation)

c Data interpretation

d Data extraction and interpretation

e to **g** Reasoning (explanation)

Level

a Easy **b** Mod **c** Mod **d** Mod **e** Mod
f Mod **g** Mod

Answers

a (i) Steak and kidney pie
(ii) Chocolate cake
(iii) Fruit yoghurt
(iv) Orange
(v) Sugar
(vi) 2120 kJ

b (i) Cornflakes, Bread, Ketchup, Chips, Baked beans, Chocolate cake, Coffee, Sugar, Chocolate, Biscuits, Banana, Orange

(ii) Protein =

Cornflakes	2.2 g
Bread	2.5 g
Ketchup	0.3 g
Chips	11.2 g
Baked beans	7.2 g
Chocolate cake	5.8 g
Coffee	2.9 g
Sugar	0
Chocolate	4.2 g
Biscuits	4.0 g
Banana	1.4 g
Orange	0.8 g
TOTAL	42.5 g

c The ratio is approximately 5:1.5:1, so there is too high a proportion of fat for the diet to be balanced.

d (i) 5350 kJ
(ii) She would become overweight.
(iii) Iron.
(iv) Reduce the high energy foods, such as chips, chocolate cake, chocolate and biscuits.

e On average their mass is greater, so more cell growth is required and more energy is expended on moving the larger mass; more energy is required to maintain body temperature owing to thinner insulating layer of fat; men have a higher rate of metabolism; men have a higher proportion of muscle tissue. Other sensible suggestions should be acceptable, but the suggestion that males are more physically active is questionable.

f Bones and teeth are still growing.

g Iron is required for haemoglobin, and females lose more iron during menstruation.

61 A hole in the stomach

Context

Digestion. The account is based, with some journalistic licence, on the reports published by Beaumont. Copies of original sections of Beaumont's papers are reprinted in several publications, including the Nuffield Biology texts. The extract could well lead to a discussion of the ethics of experimentation on human patients.

Knowledge required

a Stomach action and function.

b Human body temperature as 37 °C.

Skills

a Data extraction and interpretation

b, **c** Experimental procedure

d Data interpretation

e Experimental procedure

f Reasoning (hypotheses) and Experimental design

g Application (social)

Level

a Easy+ **b** Easy **c** Easy **d** Easy
e Mod+ **f** Mod+ **g** Easy+

Answers

a (i) Less than two hours.
(ii) Between 8 and 10 hours.
(iii) Churning effect of stomach would break up softened lumps of meat and increase surface area for digestion; constant renewal of digestive juices in the stomach might also be suggested as speeding up the process.

b To provide the same temperature conditions as in the stomach.

c Beef kept in a test-tube in water, or similar, at the same temperature, etc.

d The juice produced by the stomach is able to digest beef and has the same effect outside as inside the stomach.

e Pupils may suggest testing other meats; boiling the juice to look for evidence of enzyme activity; testing other temperatures; trying to extract the active ingredient, etc.

f (i) The spice may inhibit the production of the juice in the stomach or the action of the juice, or may coat the beef and prevent the juice getting to its surface.
(ii) Any reasonable tests relating to the suggestion made, e.g. if the main effect is on production, digestion should not be slower or prevented if the test is carried out in a test-tube. If the action of the juice is affected the rate should vary in different concentrations; coating half of a piece of meat would show a surface effect.

g Gives pupils an opportunity to consider the ethics of the experiment, and to balance the benefits of the knowledge gained against possible harm to the patient.

62 On the farm

Context

Uses of food and food production. The question also relates to energy transfer from one trophic level to another, as considered in natural ecosystems in questions 17 and 18.

Knowledge required

General understanding of energy and proteins.

Skills

a, **b** Data extraction

c Application (economic)

d Reasoning (hypotheses)

e Data extraction

f Data interpretation

g Application (economic)

Level

a Easy **b** Easy **c** Easy **d** Mod **e** Easy
f Diff **g** Mod+

Answers

a Pig.

b Cattle.

c More efficient converters should produce cheaper meat.

d E.g. more of food digested, absorbed or assimilated; less energy lost in movement or as heat; lower metabolic rate.

e Rabbit.

f Meat is not all protein. A high proportion of fat in the meat means that much of the energy is stored as fat. Rabbit and chicken both have relatively low fat contents in their meat.

g E.g. demand/marketability; labour costs; capital costs, e.g. of housing for animals; nature of land and food available, e.g. some land may be suitable for sheep grazing only; other marketable products in addition to meat, e.g. milk, wool.

63 Digestive enzymes

Context

Digestion and conditions for enzyme activity. The question refers to three of the enzymes commonly specified in GCSE syllabuses, and requires pupils to consider their properties in relation to the conditions under which they act.

Knowledge required

a to **e** Main parts of the alimentary canal; pH scale.

f Cells as having protein structure.

Skills

a to **c** Data interpretation

d Reasoning (explanation)

e, **f** Reasoning (hypotheses)

Level

a Easy **b** Easy **c** Easy **d** Mod **e** Mod
f Mod

Answers

a

Part	pH	
Mouth	6.5 to 7.5	approximately neutral
Stomach	2	acid
Small intestine	7.5 to 8	alkaline

b

Enzyme	(i) Optimum pH	(ii) Part of digestive system
A	2	stomach
B	6.8	mouth
C	8	small intestine

c A = gastric proteinase; B = amylase; C = pancreatic proteinase.

d Food only stays for a short time in the mouth. Amylase will become inactive at pH 2 in the stomach. Therefore not all starch will be digested. The rest of the starch can be digested in the neutralised contents of the small intestine.

e Strong acid conditions will kill bacteria, or other reasonable suggestion.

f Cells are made largely of protein so could be digested by the proteinase if it were active before secretion. Mucus helps to protect the lining after secretion.

64 The dangers of a sweet tooth

Context

Tooth decay. The question could be used as an extension to work on diet or micro-organisms, or as part of a health education unit.

Knowledge required

d, **e** How bacteria cause tooth decay.

Skills

a, **b** Data extraction

c Data interpretation

d Reasoning (explanation)

e Data interpretation

Level

a Easy **b** Easy **c** Easy **d** Easy **e** Easy

Answers

a 1

b 9.6 (9 to 10)

c The more sweets eaten per day, the more the number of decayed tooth found.

d Sugar tends to stick to the teeth. Pupils not familiar with microbial activity may assume the sugar has a direct effect on the teeth. Those who are aware of microbes should appreciate that the sugary layer will result in a large increase in the population of bacteria feeding on the sugar, and that acids produced by the bacteria will damage the enamel of the teeth.

e Bacteria must be needed for decay to occur, for the reasons given in part **d**.

65 Two toothpastes

Context

Teeth and dental hygiene. This could be followed up by an investigation of toothpastes along similar lines.

Knowledge required

None.

Skills

a Observation

b Measurement

c Reasoning (hypotheses)

d, **e** Application (personal)

Level

a Easy **b** Mod **c** Easy **d** Easy **e** Mod

Answers

a Particles in A are larger and more irregularly shaped.

b A – about 0.08 mm; B – about 0.04 mm.

c As an abrasive, to scrape off sticky layers of food, etc., on the teeth.

d B, because the particles are smaller, smoother and therefore less abrasive.

e Reasonable suggestions would be: a bactericide; an alkali to counteract bacterial acids; a 'creamy' base; flavouring; fluoride.

66 The fluoride problem

Context

Teeth and dental hygiene. The question relates to the public health issues surrounding the addition of fluoride to water supplies.

Knowledge required

None.

Skills

a Data interpretation

b, **c** Data extraction

d Data interpretation and Application (economic)

e to **g** Application (personal, social)

Level

a Easy **b** Easy **c** Easy **d** Mod **e** Mod+ **f** Mod+ **g** Mod+

Answers

a Number of decayed teeth decreases rapidly up to about 1 part per million; then more slowly.

b 2

c 8

d 1 to 1.5 parts per million. There is little further decrease in rate of decay above this.

e Health hazards; some people might be allergic; expense; danger of adding too much in error; possibility of some people getting extra high doses by also getting fluoride from other sources; possible ecological effects after discharge of waste water; possible effects when water used for other purposes, e.g. in cooling equipment, etc.

f By comparing rates of disease in samples of people from areas with fluoride of known concentration, the safety could be assessed. It would be important to standardise for other factors.

g E.g. in food, artificial drinks, toothpaste, pills, etc.

UNIT 9

Transport systems

67 Diffusion

Context

Rate of diffusion in relation to size of an organism. This is particularly useful as an introduction to understanding the need for a large surface area in gaseous exchange systems and the need for transport systems.

Knowledge required

The process of diffusion as movement of a substance through a medium.

Skills

a Data handling (calculation)

b Data handling (graphs)

c Data interpretation

d Reasoning (explanation)

Level

a Mod **b** Mod **c** Mod **d** Mod

Answers

a

Size of cube (mm)	Surface area (mm^2)	Volume (mm^3)	Surface-to-area-volume ratio
1	6	1	6
2	24	8	3
3	54	27	2
4	96	64	1.5
5	150	125	1.2
10	600	1000	0.6

b

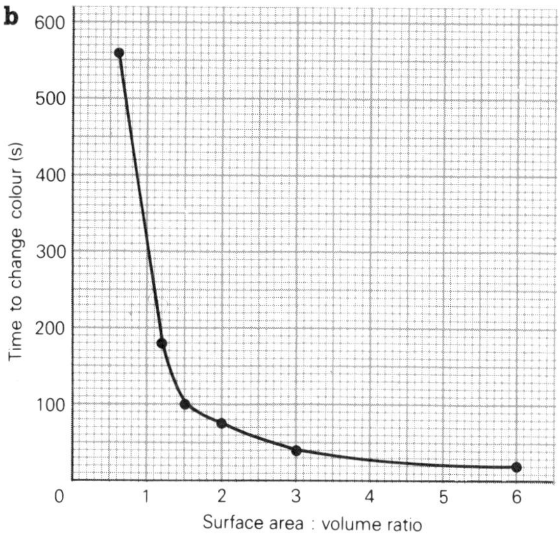

c As the ratio increases the time taken for diffusion decreases exponentially. Pupils should appreciate that the ratio decreases as the size increases.

d Being flattened, the tapeworm has a larger surface-area-to-volume ratio than the rounder earthworm. Therefore, diffusion is sufficiently rapid to meet the needs of the tapeworm, and it has no need of a transport system. With able pupils it may be worthwhile considering other factors which are relevant, such as amount of activity, and to consider other examples of familiar organisms, e.g. *Amoeba* and insects.

68 Intravenous feeding

Context

Blood and its composition. The question may be used as an illustration of an application of the principles of osmosis, and it also includes reference to blood donation and transfusion.

Knowledge required

Constituents of blood.

Skills

a Measurement and Data handling (calculation)

b Data interpretation

c Experimental procedure

d Reasoning (explanation) and Application (personal)

Level

a Easy **b** Mod **c** Mod **d** Mod

Answers

a Mean diameter is about 7 micrometres.

b The concentration at which the red cells are the same diameter as in the photograph, i.e. 0.44 g/100 cm^3.

c The most appropriate method is centrifugation, although filtration may be suggested.

d Giving whole blood means loss of blood cells which take several days to replace, but giving plasma only involves loss of water, salts and plasma proteins which are rapidly restored.

69 Counting red blood cells

Context

Blood, and requirements for the synthesis of red cells. The question concentrates on an important application of sampling and accurate counting by stratification.

Knowledge required

e Iron as a constituent of haemoglobin.

Skills

a Experimental procedures

b Observation and Measurement

c, **d** Data handling (calculation)

e Application (personal)

Level

a Mod **b** Mod **c** Mod **d** Diff **e** Mod

Answers

a Reference should be made to counting the number of cells in each small square; making allowance for cells which lie over lines; and using a tally for recording.

b

	Large Square			
	1	2	3	4
No. of blood cells	129	136	146	126

c Mean number of cells in a large square

$$= \frac{129 + 136 + 146 + 126}{4}$$

$$= 134.25$$

d Volume in large square = $0.2 \times 0.2 \times 0.1$ = 0.004 mm^3. Therefore number in 1 mm^3 of diluted blood = mean number $\times$ 250.

Dilution = 200 times. Therefore number in 1 mm^3 of undiluted blood = mean number × 250 × 200. Therefore this blood contains 6712500 red cells per mm^3.

e (i) Iron is needed for haemoglobin; therefore fewer red cells are made. (Most iron is recycled, but loss occurs due to erosion from outer epithelial layers, especially the alimentary canal, small amounts in bile, and menstruation in females.) Iron can be supplied from suitable dietary items or in tablets.

(ii) Not enough red cells made in the marrow to replace those lost. Red cells have an average life span of only about 100 days. The vitamin could be supplied in the diet, especially in meat and dairy products, or in tablets.

(iii) Even if there is plenty vitamin B_{12} in the diet it is not absorbed. Treatment would have to be by injection of the vitamin.

70 Athletes' hearts

Context

Rate of heart beat. This question could well be linked with practical work on measuring the rate of heart beat after exercise.

Knowledge required

None.

Skills

a Data handling (graphs)

b Data interpretation

c Application (personal)

d Data interpretation

Level

a Easy **b** Easy **c** Easy+ **d** Easy+

Answers

a Graphs. Weaker pupils may need assistance in selecting suitable scales.

(i)

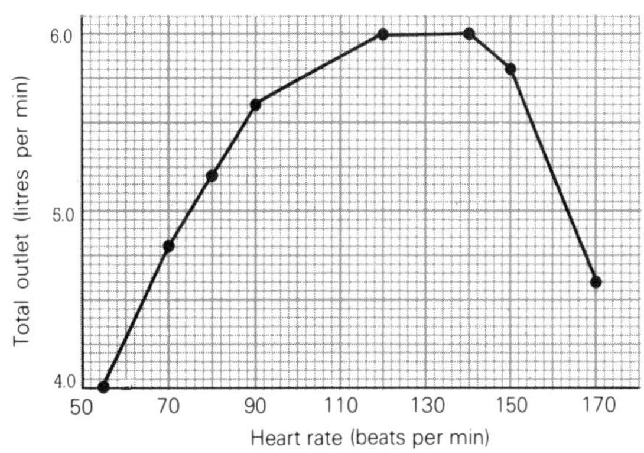

(ii)

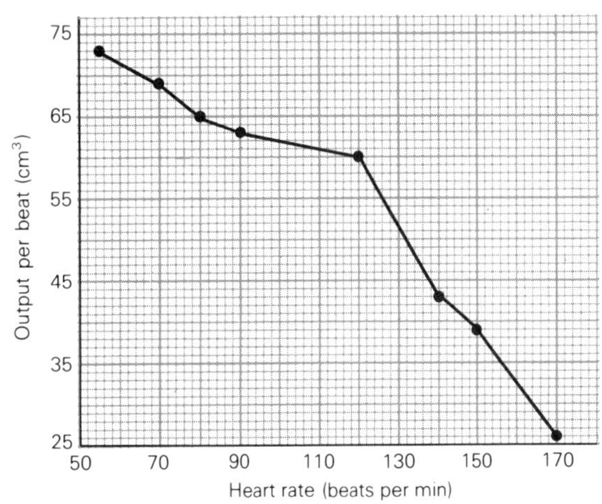

b (i) As heart rate increases the total output increases up to a maximum at 120–140 beats per minute. Above this rate the total output declines.

(ii) As heart rate increases output per beat decreases, with the rate of decline becoming greater above about 120 beats per minute.

c The heart is most efficient in terms of output at low rates of beat. If the rate of heartbeat is naturally low the increase in rate that accompanies exercise still lies in the range of maximum efficiency.

d Above this level the total output decreases, so further increase results in blood being pumped around the body less quickly and therefore poorer distribution of oxygen, glucose, etc.

71 A baby's blood system

Context

Circulation and structure of the blood system. The question requires pupils to think about the operations and functions of the circulatory system; it is not expected that they should know the arrangement of the foetal system.

Knowledge required

a to **c** Major blood vessels and chambers of the heart.

d Structure of arteries and veins.

Skills

a, b Data interpretation and Application (personal)

c Observation and Data interpretation

d Application (personal)

Level

a Mod+ **b** Mod+ **c** Mod **d** Mod+

Answers

a Most of the oxygenated blood passes directly from the placenta to the main body circulation, bypassing the lungs which have no function in the foetus.

b Blood would continue to bypass the lungs and would not therefore be fully oxygenated. The tissues of the baby would not respire efficiently, hence causing muscular weakness, low body temperature, etc.

c Umbilical vein; vena cava; right auricle (atrium); left auricle; left ventricle; aorta; artery to legs.

d Arteries have muscular walls which can contract to close the vessel. Veins have valves which prevent backflow.

72 Heart disease

Context

Blood circulation. The causes of heart disease are often required to be known by GCSE candidates, and it is assumed that these are familiar to pupils. Pupils are here asked to consider some of the processes involved in operating to treat the disease.

Knowledge required

a, b Possible causes of coronary thrombosis.

Skills

a Knowledge

b Application (personal)

c Reasoning (explanation) and Application (personal)

d Application (technological)

e Reasoning (explanation) and Application (personal)

Level

a Easy **b** Easy **c** Easy **d** Mod+
e Mod+

Answers

a Build-up of fatty deposits on the inner walls.

b Consume less fat, especially animal fats; take exercise which uses up excess fat. Pupils may suggest other measures such as not smoking and drinking less alcohol, and relaxing more.

c Cooling decreases the metabolic rate and hence the need for oxygen. In the absence of circulation this is important to keep the heart tissue alive.

d Various suggestions may be made. The blood may, as in the lungs, be passed over a thin membrane with a large surface area which is porous to gases but not liquid; clearly it would be more efficient to use oxygen rather than air.

Alternatively a form of dialysis could be used, with well-oxygenated plasma fluid. Oxygen may also be bubbled into the blood, a method which is used although there is some danger from frothing and bubbles forming in blood vessels.

e Clotting is stimulated by damage to blood vessels and also within the machine. The heparin prevents dangerous clots forming.

73 Water loss

Context

Transpiration. The question follows up the standard experiments investigating water loss from leafy shoots.

Knowledge required

g Respiration and photosynthesis.

Skills

a Data handling (graphs)

b, **c** Experimental procedure

d Reasoning (explanation)

e, **f** Reasoning (prediction)

g Knowledge

h Application (technological)

Level

a Easy **b** Easy **c** Easy **d** Easy **e** Easy **f** Easy+ **g** Easy **h** Easy+

Answers

a Graph – the two lines should be clearly labelled. See diagram opposite.

b Different shoots would have different leaf areas and could not be directly compared.

c To prevent evaporation from the water in the tube.

d By being drawn up through the shoot and evaporating from the shoot surfaces.

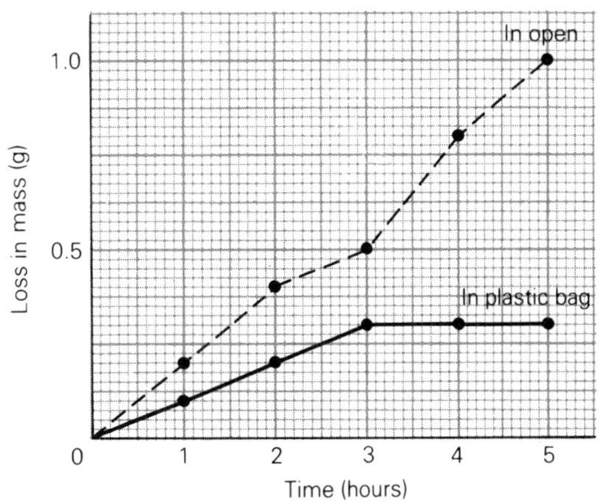

e Humidity would increase. Temperature may also rise.

f Temperature, humidity, light intensity, air currents, air pressure.

g (i) Respiration (ii) Photosynthesis

h Cover the plants with clear polythene sheeting, supported on a frame to allow entry of carbon dioxide.

74 Prize pears

Context

Translocation and the role of phloem.

Knowledge required

b, **c** Transport of nutrients through phloem.

Skills

a Experimental design

b Reasoning (explanation)

c Reasoning (prediction)

Level

a Mod+ **b** Mod **c** Mod

Answers

a Reference should be made to: using a branch of similar girth as a control (i.e. with a similar amount of phloem) and similar leaf area; ensuring that there is the same number of pears on each branch; carefully removing a ring of bark and phloem from one branch; finding the mean mass of pears on each branch at the end of the investigation.

b Sugars are no longer exported from the branch, so the pears have more materials for growth.

c The branch would be unable to import nutrients from stores in the trunk and roots for the development of new leaves and would die.

75 Mineral ion transport

Context

Transport of ions in plants. The question illustrates a use of a radioactive isotope as a tracer.

Knowledge required

Radioactive isotopes as a means of tracing elements in an organism.

Skills

a, b Experimental procedure

c, d Data extraction

e Data interpretation

Level

a Mod **b** Mod **c** Easy **d** Easy **e** Easy

Answers

a To prevent the potassium ions diffusing between the xylem and phloem.

b Any kind of paper which was permeable to ions, but stiff enough to be inserted into the shoot.

c The xylem in the separated section of the shoot with waxed paper.

d The phloem in the separated section of the shoot with waxed paper.

e Xylem, since phloem has very little potassium when it is separated from the xylem.

UNIT 10 Respiration

76 Exercise

Context

Breathing. Pupils may easily carry out similar investigations for themselves.

Knowledge required

c, **d** Relationship between breathing rate and activity.

Skills

a Data handling (graphs)

b Data interpretation

c Data interpretation and Application (personal)

d Reasoning (explanation)

Level

a Easy **b** Easy **c** Easy **d** Easy

Answers

a

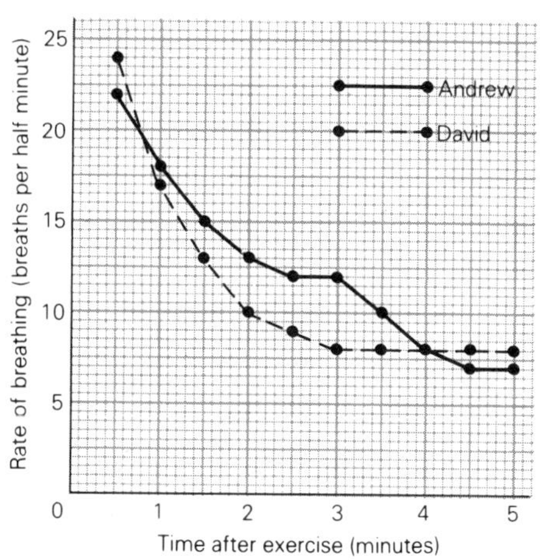

b Andrew: 4.5 minutes; David: 3 minutes.

c David. His rate of breathing returns to normal more rapidly.

d (i) It increases.
(ii) Air is brought into the lungs more rapidly, so more oxygen diffuses into the blood. Therefore more oxygen is circulated and available for respiration, which is increased during muscle activity. The rate of removal of toxic carbon dioxide is also increased.

77 Breathing

Context

Gas exchange in the lungs. The data is derived from Haldane's classic investigation of breathing carried out in the early years of the century, which contributed to an understanding of the factors involved in the control of breathing.

Knowledge required

Gas exchange in the lungs.

Skills

a, **b** Data interpretation

c, **d** Reasoning (explanation)

Level

a Easy **b** Easy **c** Mod **d** Diff

Answers

a Very little increase until percentage of carbon dioxide rises considerably, i.e. above 5.5%.

b Steady increase, with again a steep rise above 5.5%. Pupils may appreciate that this can be shown more clearly if results are plotted on a graph.

c Depth of each breath is increased.

d (i) It will be reduced, because there will be less difference in concentration and hence a reduced diffusion gradient.
(ii) The air in the lungs will be replaced more rapidly and completely; therefore more carbon dioxide is expelled and the diffusion gradient can be maintained.

78 Lung structure

Context

Gas exchange in the lungs. This question requires pupils to consider the relationship between structure and function.

Knowledge required

Basic lung structure and function.

Skills

a Measurement and Data handling (calculation)

b Data handling (calculation) and interpretation

c Experimental procedure

d Reasoning (prediction)

e Measurement

f Reasoning (explanation)

g Knowledge

h, i Reasoning (explanation) and Application (personal)

Level

a Mod **b** Diff **c** Diff **d** Mod **e** Mod
f Easy **g** Easy **h** Easy **i** Mod

Answers

a Precise answer depends on alveoli measured. Human alveoli vary from about 50 to 300 micrometres in diameter.

b (i) Pupils will use the answer obtained in part **a** for the diameter.
(ii) An alveolus is not a complete sphere, as there is an open entrance. Pupils may also point out the incomplete inflation and variability in position of section.

c Answers should be based on sampling a small part of the lungs: e.g. count the number of alveoli in several millimetre squares; take the mean; calculate the number in an average millimetre cube; find the total volume of the lungs, e.g. by displacement; make allowances for the major bronchioles, etc.; compute total internal surface area from number of alveoli. (There are said to be about 250 million alveoli in the human lungs.)

d (i) Total surface area will be reduced.
(ii) The rate of oxygen absorption will also be reduced.

e Approximately 0.13 to 0.17 micrometres.

f To allow rapid gas exchange and maintain a large diffusion gradient. Diffusion in liquids is relatively slow.

g E.g. carbon dioxide, glucose or any other nutrient, hormones, etc., but not, of course, oxygen.

h Air would be unable to enter the alveoli and therefore oxygen would have to diffuse through the water. This would be much too slow to meet the body's needs.

i During inspiration the alveoli would not expand to allow air to enter; hence a shortage of oxygen would arise. This can be a problem, especially in premature babies, who may not have started producing the surfactant necessary for them to start breathing.

79 Lungs and size

Context

Metabolic activity in mammals. Pupils need to understand that to make valid comparisons data must be expressed in standard form, i.e. per gram of body mass. The question considers factors which affect the rate of metabolic activity in mammals of different sizes and how lung surface area correlates with metabolic activity.

Knowledge required

c Heat loss in relation to size.

d Lung structure.

e Bats as flying mammals.

Skills

a, b Data interpretation

c Reasoning (explanation)

d Data handling (calculation) and Reasoning (explanation)

e Reasoning (hypotheses)

Level

a Easy **b** Mod **c** Diff **d** Mod **e** Diff

Answers

a The larger the mass of the mammal, the smaller the surface area of the lungs per gram of body mass.

b Larger mammals in general consume less oxygen per gram of body mass than smaller ones.

c A small mammal has a large surface area in relation to its volume and mass. Its rate of heat loss is thus relatively larger. Therefore its metabolic rate is higher in order to maintain body temperature, and its oxygen requirements are greater. Hence it needs a larger lung surface area in proportion to its size in order to absorb enough oxygen.

d (i) $100 \times 10 = 1000\ cm^3$
(ii) Large surface area of lungs results from the large number of branches of the bronchioles and the expansion of the ends of these to form vast numbers of alveoli.

e The most obvious suggestion is that its oxygen requirements for rapid flight are greater than those of the mouse at its most active.

80 Walking and running

Context

Energy requirements for muscular activity.

Knowledge required

e Changes in breathing and circulation during exercise.

Skills

a Data extraction

b Data interpretation

c Reasoning (explanation)

d Data interpretation

e Reasoning (hypotheses)

Level

a Easy **b** Easy **c** Easy+ **d** Mod **e** Diff

Answers

a (i) 0.8 litres per kg per hour.
(ii) 3.2 litres per kg per hour.
(iii) 0.25 litres per kg per hour.

b (i) The faster a man runs the greater the oxygen consumption.
(ii) Muscular activity requires oxygen for respiration. Therefore the more muscular activity, the more oxygen is consumed.

c Increased rate of breathing; increased depth of breathing; increased rate of heartbeat/circulation; dilation of blood vessels in muscles.

d About 8 kg per hour.

e Running involves longer strides and therefore the mass of the body has to be lifted fewer times in a given distance. Placing the foot flat on the ground at each step in walking may also require some muscular activity, and it may be argued results in more friction.

81 Running, swimming and flying

Context

Movement and energy. The question considers the ways in which organisms are adapted for movement in different media.

Knowledge required

f Basic knowledge of fish structure.

h, i Migration.

Skills

a, b Data extraction

c, d Data handling (calculation)

e, f Reasoning (inference and explanation)

g Reasoning (hypotheses)

h Application

i Reasoning (hypotheses)

Level

a Easy **b** Easy **c** Mod **d** Mod **e** Easy
f Easy+ **g** Diff **h** Diff **i** Diff

Answers

a Swimming.

b 0.9 kJ

c $\frac{36}{1.5} = 24$ km

d Just over 4 g (4.17 g).

e Air. Air is less dense and therefore offers less resistance.

f Streamlined shape; smooth surface; powerful tail muscles; large surface area of caudal fin.

g The mass of the body has to be lifted off the ground at each step. Much energy is therefore used for vertical movement rather than horizontal movement. Each limb is also accelerated and then decelerated again as it regains position. Some energy is used to overcome air resistance, but this is a relatively small proportion in small mammals.

h Mass of bird; distance to be flown; possibility of feeding *en route*. Wind resistance, rate of heat loss, flying efficiency, etc., might also be relevant.

i Flying uses much less energy per kilometre than running. It is also much more rapid. The energy cost and time involved for a small mammal would not be balanced by improved feeding opportunities, etc. Physical barriers also make land migration more difficult.

82 Diving ducks

Context

Breathing and circulation. The data show an example of physiological adaptation.

Knowledge required

c, d Relationship between breathing and circulation.

Skills

a Data extraction

b Data handling (calculation)

c Reasoning (hypotheses)

d Data interpretation and reasoning (inference)

Level

a Easy **b** Mod **c** Mod **d** Mod+

Answers

a From a steady rate of heartbeat while swimming on the surface, the rate increases when preparing to dive. The rate decreases again while the duck is underwater and returns to a high rate after resurfacing.

b (i) Approximately 320 beats per minute.
(ii) Approximately 150 beats per minute.

c The duck needs to supply as much oxygen as possible to its muscles before submerging.

d The rate of breathing might be expected to be greater during period B than in A. Breathing would then cease during period C as it dives underwater.

83 Hibernation

Context

Respiration and maintenance of body temperature.

Knowledge required

Relationships between breathing, heart rate, respiration and temperature control.

Skills

a, **b** Data extraction

c to **e** Reasoning (prediction)

f to **i** Reasoning (explanation) and Application.

Level

a Easy **b** Easy **c** Easy **d** Easy **e** Easy
f Easy **g** Easy **h** Easy **i** Easy

Answers

a Decreases to 5 °C.

b Both decrease.

c Decreases.

d Decreases.

e Decreases.

f Rate of respiration decreases, so less oxygen is needed.

g Less oxygen and glucose need to be circulated.

h As a source of energy with a high energy content per gram; as an insulating layer to reduce heat loss.

i To help maintain heat, since a smaller overall surface area is exposed.

84 Wine

Context

Anaerobic respiration. The question deals with the application of yeast's ability to respire anaerobically and produce alcohol.

Knowledge required

f, **g**, **h** Anaerobic respiration in yeast.

Skills

a Data interpretation

b to **e** Data handling (translation)

f to **h** Reasoning (explanation) and Application (technological)

Level

a Easy+ **b** Easy **c** Easy **d** Easy
e Easy **f** Easy **g** Mod **h** Mod

Answers

a The flow chart should show the following stages, and may be suitably illustrated:

pick grapes → crush grapes → must → treat with sulphur dioxide → fermentation → liquor → filter → bottle

b Yeast growing on the grape skins.

c To prevent growth of unwanted bacteria and yeasts.

d 15%

e Either by using must with a very high sugar content, or by stopping fermentation before all sugar is used up.

f If oxygen is present aerobic respiration occurs, and complete oxidation to carbon dioxide and water takes place.

g The carbon dioxide produced remains in the bottle, largely in solution under pressure. When the pressure is released the carbon dioxide comes out of solution and the sparkling bubbles are produced.

h The late season grapes contain large amounts of sugar. Not all of this sugar is used up during fermentation.

UNIT 11 Keeping a balance

85 Camels in the desert

Context

Temperature control and water conservation. The question also illustrates adaptation to environmental conditions.

Knowledge required

b Sweating.

Skills

a Data interpretation

b, **c** Reasoning (explanation) and Application

Level

a Easy+ **b** Easy **c** Mod

Answers

a The graph should indicate 34 °C at 0600 h. The temperature should rise at a steady rate until 1800 h, when it should be 40 °C. The temperature should then fall back to 34 °C by 0600 h.

b Sweating involves the loss of water, which the camel needs to conserve.

c A low starting temperature means that it will take longer for the body temperature to rise to 40 °C, and thus less sweating and water loss will occur.

86 A desert plant

Context

Adaptation to environmental conditions. The question provides an unfamiliar example of the cooling effect of evaporation comparable to that achieved by sweating.

Knowledge required

Stomata. Cooling effect of evaporation.

Skills

a, **b** Reasoning (explanation)

Level

a Mod **b** Mod

Answers

a Opening of the stomata allows water to evaporate, and evaporation of water needs heat. The latent heat of vaporisation is taken mainly from the leaves.

b Loss of water, which may be difficult to replace in a desert area.

87 Spiny anteaters

Context

Mechanisms of temperature control. Pupils are asked to apply their understanding to an unfamiliar example.

Knowledge required

Mechanisms of temperature control in a mammal.

Skills

a Data interpretation and Reasoning (inference)

b Reasoning (hypotheses)

Levels

a Easy **b** Easy+

Answers

a (i) The body temperature might be expected to be much the same as the environmental temperature, or at least closer to it than in mammals with the temperature control mechanisms listed. Activity would generate heat which might be retained and raise the temperature in the colder conditions.
(ii) It would remain constant at about 37 °C, except perhaps at the highest temperatures when it might rise above this.

b Behavioural adaptations, such as burrowing, are the most feasible suggestions.

88 Kidney machines

Context

Kidney function. Understanding of the processes which take place in the kidney is related to the design of kidney machines. Teachers may wish to extend discussion to consideration of the issues concerned with provision of kidney machines and kidney transplantation.

Knowledge required

Basic kidney function.

Skills

a, **b** Reasoning (explanation)

c Reasoning (prediction)

d Application (medical)

e Application (technological)

Level

a Mod **b** Mod **c** Mod **d** Mod **e** Mod

Answers

a Protein molecules are too large to pass out of the blood plasma into the kidney tubules.

b All the glucose is re-absorbed from the kidney tubules into the blood plasma.

c It would have higher concentrations of urea and mineral ions.

d (i) Urea.
(ii) Protein and glucose.
(iii) Water, and sodium and chloride ions.

e The same composition as blood plasma with respect to water, glucose and ions, but with no urea or other nitrogenous waste materials.

89 A crab's problem

Context

Osmo-regulation. Changing environmental conditions can alter the osmotic potential of a rock pool, and therefore organisms like crabs which live there have to be adapted to cope with this.

Knowledge required

c Principles of osmosis.

e Active transport.

Skills

a, **b** Reasoning (hypotheses)

c Reasoning (explanation)

d Data interpretation

e Reasoning (hypotheses)

Level

a Mod **b** Mod **c** Mod **d** Easy **e** Diff

Answers

a Heat causing evaporation of water; or flow of sea water into a diluted pool.

b Rain.

c The solution outside would become less concentrated than the fluids in the crab, and therefore water would tend to enter the crab by osmosis.

d It increases.

e An increase in oxygen consumption suggests an increase in respiration, and this might well be used to pump out the excess water by active transport.

90 Diabetics

Context

Regulation of blood sugar concentration.

Knowledge required

e Insulin and its effect.

Skills

a Experimental procedure

b, **c** Data extraction

d Data interpretation

e, **f** Application (medical)

Level

a Easy **b** Easy **c** Easy **d** Easy **e** Easy **f** Mod

Answers

a To make sure there is minimal glucose already in the blood.

b 350 mg per 100 cm^3.

c 1 hour 45 minutes.

d About 2 hours 30 minutes.

e The curve would be similar to that for the healthy person.

f Hormones are broken down in the liver. Remaining for long periods in the blood would negate their usefulness as control agents.

91 An aquatic toad

Context

Excretion of nitrogenous waste. *Xenopus* is able to excrete ammonia while in water but converts it to urea when removed from the water.

Knowledge required

Urea is formed from ammonia.

Skills

a Data handling (graphs)

b, **c** Data interpretation

Level

a Mod **b** Diff **c** Diff

Answers

a

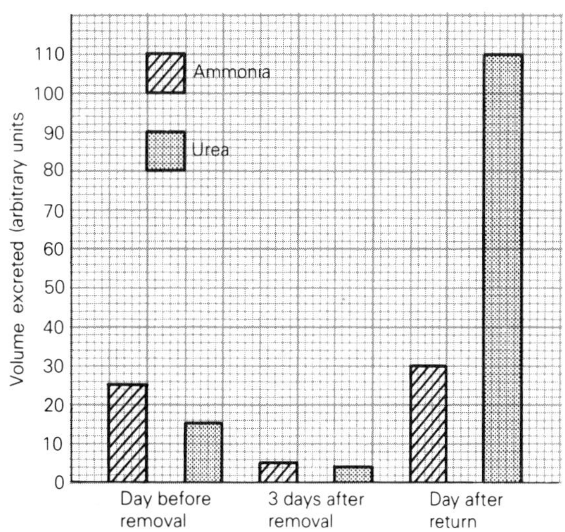

b Ammonia and urea are normally excreted in solution. When out of water *Xenopus* is unable to excrete large amounts of water, and therefore the substances are retained.

c Much of the ammonia was converted to urea which is less toxic. This was then flushed out as soon as the *Xenopus* returned to the water.

UNIT 12 Control and coordination

92 Reaction times

Context

Nervous control. The 'reaction time' represents the time taken for a nerve impulse to pass from the eye via the brain and spinal cord to the relevant muscles, and therefore demonstrates the relative slowness of nervous conduction.

Knowledge required

None.

Skills

a Data handling (tabulation)

b Data handling (graphs)

c Data extraction

d to **f** Data handling (calculation)

Level

a Easy **b** Easy **c** Easy **d** Easy **e** Easy
f Mod

Answers

a

Number of pupils	Reaction time (hundreths of a second)
15–19	6
20–24	9
25–29	9
30–34	4
35–39	2

Pupils should be encouraged to use a tally to find the number in each frequency class.

b

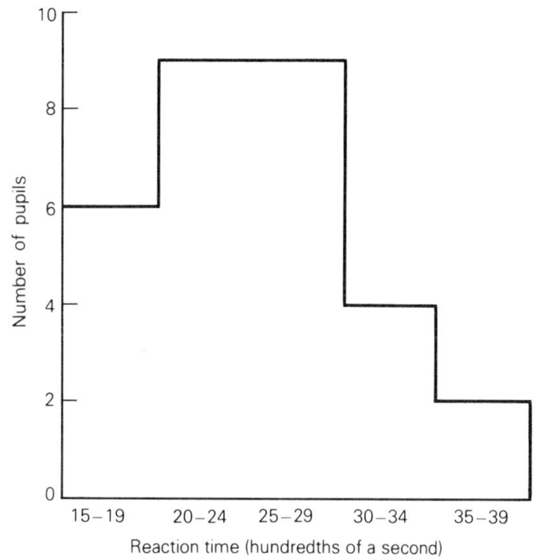

c (i) Pupil 3. (ii) Pupil 14.

d 25.03 hundredths of a second.

e 50%

f (i) 2.50 m (ii) 6.17 m

93 Eyes

Context

Accommodation in the eye.

Knowledge required

Mechanism of accommodation in the human eye.

Skills

a Reasoning (explanation)

b, **c** Data interpretation

d Reasoning (prediction) and Application (personal)

e Data interpretation

f Reasoning (application) and Application (personal)

g Experimental design

Level

a Easy+ **b** Mod **c** Mod **d** Mod **e** Mod **f** Diff **g** Mod+

Answers

a Changing the shape of the lens permits the eye to focus to different distances. A thicker lens refracts light more and enables the eye to focus on closer objects.

b (i) About 56.
(ii) The person will be unable to accommodate and therefore will be unable to focus on close objects.

c The 10-year-old will be able to see objects clearly rather nearer the eye than the 20-year-old. (The average 10-year-old has a near point of about 7 cm; by the age of 20 the near point will be about 10 cm.)

d Convex converging lenses, as a thin lens is unable to refract light from near objects sufficiently.

e No further change in elasticity of the lens occurs beyond this age.

f Much of the refraction and image formation results from the curvature of the cornea. For distant vision this is sufficient for an image to be produced without the additional adjustment of the lens.

g Investigations should include a suitable control, e.g. using cards with spots 2 mm in diameter and blank cards, or with lines 2 mm apart and with no gap. Having prepared the material the investigator should: place the person with the head in a fixed position; ensure good lighting; hold cards at various distances from eye; randomise order of cards; record responses at different distances; repeat several times.

94 The thyroid gland

Context

Hormonal control in mammals. The question provides an illustration of a medical application for radioactive isotopes.

Knowledge required

Metabolism. Radioactivity and its measurement.

Skills

a to **c** Data interpretation

d Experimental procedure

e Reasoning (prediction)

f Reasoning (explanation) and Application (medical)

Level

a Mod **b** Mod **c** Mod+ **d** Mod **e** Diff **f** Diff

Answers

a C

b A

c Since the metabolic rate is low the person is likely to be slow-moving; lethargic; liable to feel cold and have a low body temperature; liable to put on fat since stores will be used slowly; stunted in growth when young and slow to repair tissue damage when adult.

d By means of a Geiger–Müller tube and ratemeter.

e E.g. relatively weak source to render it comparatively harmless; able to produce detectable radiation; short half-life so that radioactivity does not persist; behaves in the same way as a non-radioactive isotope of the same element.

f There would be an excess of non-radioactive iodine available for uptake by the thyroid, so only a small proportion of the radioactive iodine would be absorbed.

95 Phototropism

Context

Sensitivity in plants. This question extends the conventional work on the responses of shoots to light.

Knowledge required

Phototropism.

Skills

a, b Data interpretation

c Reasoning (explanation)

d Experimental design

Level

a Easy **b** Mod **c** Mod **d** Mod+

Answers

a (i) 440 nm
(ii) Indigo blue.

b (i) The shoot would bend towards the light.
(ii) The shoot would grow straight up, i.e. the light would not cause bending.

c It ensures that shoots grow into optimum light intensity for photosynthesis, and that leaves face the light.

d Set up lights with different colour filters to provide various wavelengths, but equal intensity; have large numbers of shoots in each colour; arrange light to shine from one side; leave for given length of time; measure angle of curvature; find average angle of curvature in each wavelength.

96 Geotropism

Context

Sensitivity in plants.

Knowledge required

b The effect of light on auxin.

Skills

a Experimental design

b Reasoning (hypotheses)

Level

a Mod+ **b** Diff

Answers

a Germinate two batches of seedlings between sheets of damp blotting paper; remove top sheet; keep one batch in light and one in dark; after fixed and equal length of time, measure angle between shoot stem and horizontal in each batch, or count number in vertical position. An alternative strategy is to use two batches of seedlings with straight shoots and rotate them so that the shoots are horizontal, measuring the time taken for the shoots to grow back to a vertical position.

b Light inactivates auxin, and therefore in the shoots kept in the light there is less auxin to elicit the growth response.

97 Sticklebacks

Context

Stimulus and response. The question illustrates how the behaviour of an animal can be adapted so that particular stimuli elicit beneficial responses.

Knowledge required

Gas exchange and respiration.

Skills

a Reasoning (hypotheses)

b Data extraction

c Reasoning (hypotheses)

d Data interpretation and Reasoning (prediction)

Level

a Easy **b** Easy **c** Mod **d** Mod

Answers

a To provide a supply of oxygenated water, and help to remove carbon dioxide.

b Day 8.

c Low oxygen concentration in water; or, high carbon dioxide concentration in water.

d (i) Day 15.
(ii) Graph should show a rise to day 6, then a sharp fall and subsequent rise to day 14, followed by a second steep fall.

98 Waterfleas

Context

Stimulus and response. Pupils are here being asked to appreciate that organisms may respond to stimuli other than those that human beings can detect.

Knowledge required

Oxygen as a product of photosynthesis.

Skills

a, b Reasoning (hypotheses)

Level

a Mod+ **b** Mod+

Answers

a Cover or protection from predation; high oxygen concentration; or other reasonable suggestions.

b They may be able to detect the high oxygen concentration, or low carbon dioxide concentration.

UNIT 13 Growth and reproduction

99 Growing up

Context

Factors affecting growth in human beings. The first parts relate to rate of growth through childhood, the remainder to the effects of metal ions on embryonic development.

Knowledge required

d Factors affecting growth.

Skills

a, **b** Data extraction

c Data handling (graphs)

c Reasoning (prediction)

e to **h** Data interpretation

Level

a Easy **b** Easy **c** Mod **d** Easy **e** Easy
f Mod **g** Easy **h** Mod

Answers

a (i) 2 kg (ii) 8 kg (assuming steady fall in rate from 10 to 6 kg per year).

b About 5 months.

c The sketch (top right) should show rapid growth in the first year, then a steady but lower rate until there is a sudden increase between 13 and 15. The rate then falls almost to zero by 18.

d Diet; amount of activity; disease; genetic constitution, etc.

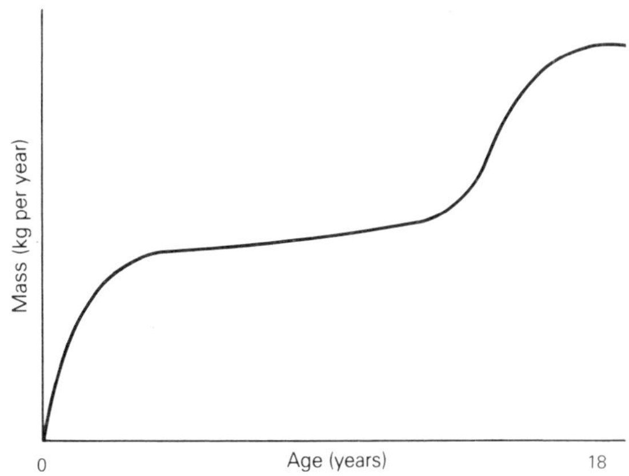

e Gain in height will cease at about 18, so the graph would drop to zero. There may be continuing gain in mass beyond this age, and the curve may fluctuate, probably showing some gain around middle age and a fall in old age.

f Increasing concentration of cadmium and lead both result in decreasing birth mass. Increasing concentration of zinc correlates with increasing birth mass up to about 60 micrograms per gram.

g Cadmium and lead.

h High concentrations of lead decrease birth mass so exposure to exhaust fumes, e.g. in urban areas, is likely to be damaging. Heavy smoking is also likely to be damaging. The effects of cadmium and lead may be enhanced by preventing sufficient zinc being absorbed to meet the needs of healthy development.

100 Ageing

Context

Pupils are asked to consider some of the changes which occur as human beings get older. The data are derived from a study of the participants in a cross-country race in Sweden.

Knowledge required

None.

Skills

a Data handling (graphs)

b Data extraction

c Data interpretation

d Reasoning (hypotheses)

e Application (social)

Level

a Easy **b** Easy **c** Mod **d** Mod **e** Easy+

Answers

a

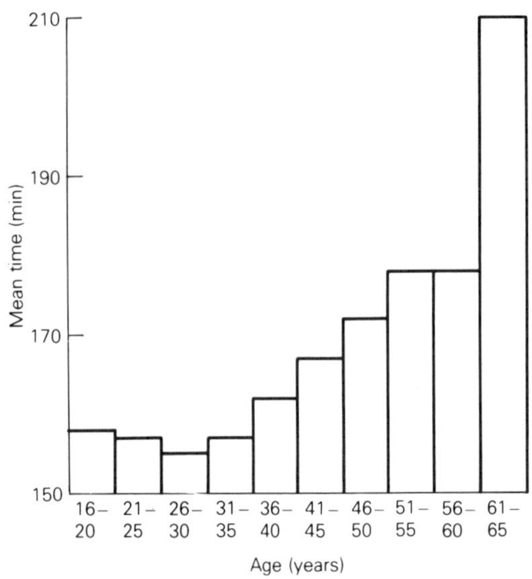

b 26 to 30

c (i) 56 to 60-year-olds might be expected to have a higher mean time.
(ii) Probably small numbers of individuals taking part in the race.

d Any reasonable suggestion, such as reduction in amount of muscle; reduced metabolic rate; reduced elasticity or ability of muscle to contract; reduced heart or lung efficiency; reduced blood flow due to narrowing of vessels.

e Increased proportion of non-working members of the population; increased demands on health care and related services; increased demands for food, housing, etc; problem of pension provision; possible social problems arising from reduced mobility, family tensions, etc.

101 Cell division

Context

Mitosis. It is not expected that pupils should be familiar with the names of the stages of mitosis nor that they should know the cytological details of the process. Part **d** is primarily included as an observation exercise; if a suitable film is available it might be coupled with an explanation of the dynamics of the process, thus making the appearance of different arrangements in slides or photographs more comprehensible.

Knowledge required

None.

Skills

a Observation

b Data handling (calculation)

c Data interpretation

d Observation

Level

a Easy **b** Easy **c** Mod **d** Diff

Answers

a 4

b 18%

c i About ¼ to ⅓

ii A much higher proportion than in a palisade cell

d Interphase, prophase, metaphase, early anaphase, late anaphase, telophase

102 Flowers

Context

Insect pollination in flowers.

Knowledge required

Main parts of a flower. Pollination.

Skills

a Observation

b, **c** Data interpretation

d Reasoning (explanation)

e Observation

f Data interpretation

Level

a Easy **b** Easy **c** Mod **d** Mod **e** Easy **f** Mod

Answers

a Anthers at top of 'petal-tube' in thrum-eyed, half-way down in pin-eyed.
Stigma at top of 'petal-tube' in pin-eyed, half-way down in thrum-eyed; or long style in pin-eyed, short style in thrum-eyed.

b E.g. large petals; petals formed into landing stage; anthers not exposed; stigmas not exposed; presence of nectary. Pupils should avoid including features not visible in the diagrams, such as colour of petals or scent.

c Bee settles on petals and extends tongue down to nectary; pollen from anthers sticks to middle part of the tongue; bee flies to a thrum-eyed flower and again extends tongue down to nectary; pollen collected on the middle part of the tongue is transferred to the stigma.

d Cross-pollination is much more likely. Cross-pollination maintains variation and hence the potential for survival in varying conditions.

e One type of flower (left-hand diagram) has male parts only; the other female parts only. The male flower has stamens and only vestiges of the female part; the female flower has extended stigma/styles and an ovary with many ovules. In the female flower petals appear to be larger, and the sepals bulge out much more.

f Cross-pollination is the only possible method as the pollen is produced in different flowers from the ovules.

103 Human chromosomes

Context

Meiosis and chromosome pairs. The question refers to the abnormal presence of an extra chromosome as in Down's syndrome, and can be used to illustrate how congenital disorders can arise.

Knowledge required

Chromosomes arranged in homologous pairs in adult cells – 23 pairs in human beings.

Skills

a Observation

b Data interpretation and Reasoning (hypothesis)

c, **d** Observation

Level

a Easy **b** Mod+ **c** Easy **d** Mod

Answers

a 47

b (i) There is an odd number of chromosomes. There is one chromosome extra to the normal human complement; therefore one chromosome must be unpaired, or one pair has an extra duplicate.
(ii) The most likely explanation is that one pair of chromosomes failed to separate in

meiosis so that a gamete contained two copies of the same chromosome.

c (i) Male.
(ii) X chromosome is larger, has more bands and an obvious 'neck'.

d A and C; B and E; D and F.

104 Population growth

Context

Human populations. Pupils should appreciate that change in population size is not simply a factor of birth rate.

Knowledge required

None.

Skills

a to **c** Data interpretation and Reasoning (prediction)

Level

a Mod **b** Mod **c** Mod

Answers

The age pyramids should show the following features:

a large numbers in the younger categories;

b relatively small numbers of people in the youngest category and larger numbers of old people;

c largest numbers in the middle category.

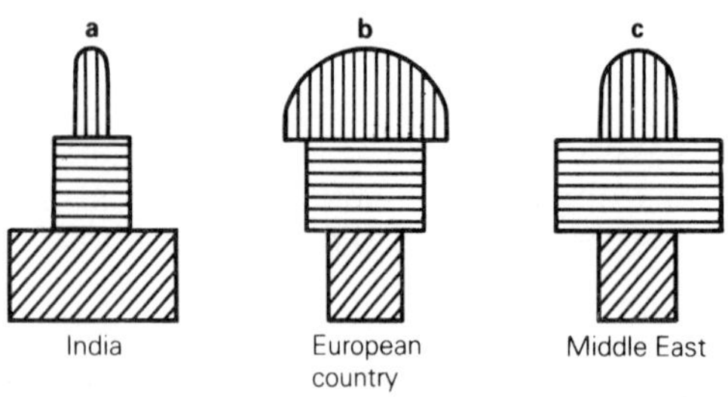

105 Contraception

Context

Population growth and control. The question offers comparative data on the effectiveness of different methods of contraception and requires no value judgements. Pupils are, however, invited to consider the value judgements that may be applied in selecting a method of contraception.

Knowledge required

Familiarity with the various methods will make the question more meaningful.

Skills

a to **c** Data extraction

d Data interpretation

e Application (social)

Level

a Easy **b** Easy **c** Easy **d** Easy+
e Easy+

Answers

a Contraceptive pill.

b Withdrawal.

c The pregnancy rate is much reduced, i.e. effectiveness as a method of contraception is much improved.

d There is no precise definition of the safe period. Individuals differ in terms of the length of the oestrous cycle and time of ovulation, and there may be variability from one cycle to the next.

e Such factors as religious or conscientious objections or taboos; cost; availability; need for medical or professional oversight; possibility of side effects; ease of use; aesthetic factors relating to use.

106 Gestation

Context

Development. The question provides practice in handling a relatively large amount of data.

Knowledge required

None.

Skills

a Data extraction and handling

b Data interpretation

c Reasoning (hypotheses)

d, **e** Data interpretation

Level

a Easy **b** Easy+ **c** Mod **d** Mod **e** Mod+

Answers

a

Mammal	Mass (kg)	Gestation time (days)
Mouse	0.025	21
Rat	0.5	22
Guinea pig	0.8	60
Hedgehog	0.8	60
Rabbit	1.5	30
Cat	4	60
Badger	15	180
Chimpanzee	75	270
Pig	300	115
Horse	1300	335
Elephant	7000	640
Whale (Blue)	120000	330

Note: the masses shown generally represent the maximum mass reached.

b (i) Not supported. There are several exceptions, e.g. a whale has a shorter gestation time than an elephant, and a rabbit than a guinea pig or hedgehog.

(ii) Supported by the only example available, i.e. the guinea pig and hedgehog. Clearly pupils should not conclude that the statement is universally true.

(iii) Not supported. Pigs, for example, have larger litters than much smaller mammals.

(iv) Supported in general, although there is not a precise relationship.

(v) The two largest do have only one young per litter, but horses may have two.

c E.g. diet, climate, size of mammal, stage of development at birth (which relate to habitat, nature of parental care, whether herbivores or carnivores, etc.).

d (i) It would be reasonable to compare it with the rabbit, which is related and has similar habits, and the cat, which is a similar size, and estimate between 40 and 70 days. The average gestation time is 42 days.

(ii) The lion might be expected to be over 100 days, but much less than the chimpanzee which only has one young per litter. Since the lion is a hunter which would be encumbered by carrying a large mass of a foetus, a relatively short gestation period could be anticipated. The average is 106 days.

e (i)

Number of babies born	Total mass of babies (kg)
Single	3.4
Twins	4.8
Triplets	5.4
Quads	5.6

(ii) The larger number and mass of multiple foetuses will fill the uterus sooner and increase demands on food supply, etc. It may also be suggested that stretching of the uterus triggers birth, or similar.

107 Eggs

Context

Sexual reproduction and fertilisation. The question illustrates adaptations for reproduction in different groups of vertebrates, and pupils may need access to information relating to the organisms mentioned.

Knowledge required

Internal fertilisation and development in vertebrate groups.

Skills

a, **b** Knowledge

c, **d** Reasoning (explanation)

e Reasoning (prediction)

Level

a Easy **b** Easy **c** Easy+ **d** Easy+
e Mod

Answers

a Human being, mouse, swan, sparrow, turtle.

b Human being and mouse.

c Where development occurs externally the egg has to include all the stored nutrients necessary until the organism can take in food. The eggs of organisms with internal development therefore tend to be smaller. Size at hatching is also relevant and organisms which hatch on land tend to be larger since very small organisms dry up rapidly and have more difficulty obtaining food.

d It depends on the likelihood of survival, which in turn depends on the number of predators and degree of parental or other protection.

e Two. For the numbers to remain stable the parents must simply replace themselves.

UNIT 14

Inheritance and evolution

108 Fingerprints

Context

Variation.

Knowledge required

c, **e** Basic understanding of genes and inheritance.

Skills

a, **b** Observation

c Reasoning (prediction)

d Observation

e Data interpretation

Level

a Easy **b** Easy **c** Mod **d** Easy **e** Mod

Answers

a Matthew – arch
Jayne – loop
Richard – arch
Andrew – loop
John – arch
Peter – loop
Michael – arch
Graham – loop
Robert – arch
Howard – whorl
Mary – loop
Louise – loop

b Robert.

c They would be identical in identical twins, but not necessarily in other twins.

d No.

e Fingerprints cannot be simply the result of the effect of genes. Even in genetically identical individuals they develop differently.

109 Tobacco plants

Context

Monohybrid inheritance.

Knowledge required

c to **e** Principles of monohybrid inheritance.

f Function of chlorophyll.

Skills

a Observation

b Data handling (calculation)

c to **e** Data interpretation

f Reasoning (explanation)

Level

a Easy **b** Easy **c** Easy **d** Easy **e** Mod
f Mod

Answers

a

Seedlings	Number
Normal	87
Albino	28

b 3:1

c *cc*

d Both *Cc*, as both must be carrying the recessive gene for albinism.

Parental phenotypes	normal	normal
Parental genotypes	*Cc*	*Cc*
Gametes	*C* or *c*	*C* or *c*
Offspring genotypes	*CC* *Cc* *Cc*	*cc*
Offspring phenotypes	normal	albino

e About one third.

f Having no chlorophyll, they would be unable to photosynthesise and grow. They will therefore die once food reserves are used up.

110 Maize

Context

Monohybrid inheritance. It may be necessary to stress to pupils that each grain is in effect an offspring.

Knowledge required

a, **d** Principles of monohybrid inheritance.

Skills

a Data interpretation

b Observation

c Data handling (calculation)

d Data interpretation

Level

a Easy **b** Mod **c** Mod **d** Mod

Answers

a Purple.

b

	Row				Total
	A	B	C	D	
Purple					
Yellow	17	10	12	10	49

c 3:1

d

Parental phenotypes	purple	purple
Parental genotypes	*Pp*	*Pp*
Gametes	*P* or *p*	*P* or *p*
Offspring genotypes	*PP* *Pp* *Pp*	*pp*
Offspring phenotypes	purple	yellow

111 Oil palms

Context

Monohybrid inheritance with codominance, and plant breeding. The question illustrates an important application of a genetic breeding programme and modern developments in cloning technques.

Knowledge required

Principles of monohybrid inheritance.

Skills

a Data interpretation

b Reasoning (explanation)

c Application (economic)

d Data interpretation

e Application (economic)

Level

a Easy **b** Easy **c** Easy+ **d** Mod **e** Mod+

Answers

a (i) *SS* (ii) *ss* (iii) *Ss*

b Take pollen from a shell-less palm and brush it onto the stigmas of a thick-shelled palm. Take precautions to prevent any other fertilisation by removing anthers of thick-shelled palm flowers and covering the artificially pollinated flower. Collect and plant seeds.

c There is a thicker fleshy layer from which the oil is obtained. Therefore each fruit will yield more oil and more profit to the grower.

d Half.

e From cross-breeding, only half of the offspring will be the economically desirable thin-walled type. Cloning will give 100% of this type. In growing new plantations, the grower can therefore be certain that all the palm trees will be profitable.

112 Blood groups

Context

Monohybrid inheritance with codominance. Blood groups provide one of the relatively small number of human examples which illustrate simple monohybrid inheritance. Nevertheless the results of a classic survey by Wiener shows anomalies.

Knowledge required

Principles of monohybrid inheritance.

Skills

a, b Data interpretation

c Reasoning (hypotheses)

d Data interpretation

e Data handling (calculation) and interpretation

f Application (personal)

Level

a Easy **b** Easy **c** Diff **d** Mod **e** Mod
f Mod

Answers

a The one *MN* child from the *M* × *M* parents, and the one *N* child from the *M* × *N* parents.

b *M* × *M* should yield children of blood group *M* only, and *M* × *N* children of blood group *MN* only.

c The exceptions may be children where the father was not the parent presumed by the researchers; there might have been a mutation; there might have been a fault in development so the gene was not expressed normally.

d

	(i)	(ii)	(iii)
Parental phenotypes	*M* × *MN*	*N* × *MN*	*MN* × *MN*
Parental genotypes	*MM* × *MN*	*NN* × *MN*	*MN* × *MN*
Gamete	*M* *M* or *N*	*N* *M* or *N*	*M* or *N* *M* or *N*
Offspring genotypes	*MM* *MN*	*NN* *MN*	*MM* *MN* *MN* *NN*
Offspring phenotypes	*M* *MN*	*N* *MN*	*M* *MN* *MN* *NN*
Ratio	1 : 1	1 : 1	1 : 2 : 1

e (i) $M \times MN = 1 : 1.07$
$N \times MN = 1 : 1.07$
$MN \times MN = 1 : 1.99 : 0.89$

(ii) These results are close to the expected ratios and do therefore support the genetic theory.

f (i) It is impossible to say on the basis of this evidence. The father of the child would be either *N* or *MN*. The partner could therefore have produced a child with the blood group *N*.

(ii) Check whether other blood groups, or characteristics, also match the partner's. Some pupils may also suggest 'genetic fingerprinting' techniques.

113 Ladybirds

Context

Discontinuous variation and natural selection. In this example, both forms of a gene convey advantages, and the overall benefits balance each other, so both forms persist in the population.

Knowledge required

i Principles of monohybrid inheritance.

Skills

a to **d** Data extraction

e, **f** Data interpretation

g Data extraction and Reasoning (hypotheses)

h Reasoning (hypotheses)

i, **j** Data interpretation

k Reasoning (application)

Level

a Easy **b** Easy **c** Easy **d** Easy **e** Easy
f Easy **g** Mod **h** Diff **i** Mod **j** Mod
k Diff

Answers

a 29%

b 71%

c 38%

d 32%

e The proportion of black ladybirds increases between spring and autumn and then decreases again during hibernation.

f Year 2.

g (i) Red.

(ii) Any reasonable suggestion should be acceptable, e.g. fewer red ladybirds are eaten during hibernation because of the red warning colour; black ladybirds lose more heat by radiation in winter and thus use up food stores faster; the black colour is linked with some other character which is at a disadvantage in winter.

h Again any reasonable suggestion, e.g. black forms have better camouflage; gain heat more rapidly; produce more young. (Research suggests that heat absorption is in fact the major factor.)

i A black individual could be either homozygous or heterozygous. The red ladybird must be homozygous recessive.

Parental genotypes	*BB* × *bb*	*Bb* × *bb*	
Gametes	*B* *b*	*B* or *b*	*b*
Offspring genotypes	*Bb*	*Bb*	*bb*
Offspring phenotypes	black	black	red

j Either 100% black, or 50% black and 50% red.

k Although at a disadvantage in the summer, it is more likely to survive hibernation. The gene for redness is maintained in heterozygotes in the population.

114 The peppered moth

Context

Natural selection. The question provides detailed data relating to the classic example of industrial melanism in the peppered moth.

Knowledge required

None.

Skills

a, **b** Data interpretation

c Data extraction

d Data interpretation

e, **f** Reasoning (hypotheses)

g Experimental procedure

Level

a Mod+ **b** Mod+ **c** Easy **d** Easy
e Mod **f** Diff **g** Mod

Answers

a (i) Untrue. Three collections in the north-east were all black.
(ii) Certainly true.
(iii) Untrue. In a couple of sites in the south there were darkish moths but no black moths.
(iv) Possibly true. None was collected, but they could be present.
(v) Possibly true. There were only three collection points, and there is no evidence relating to the rest of Scotland.
(vi) Possibly true. There is no evidence relating to temperature.
(vii) Possibly true. This is generally supported by the evidence except for the south-east.
(viii) Possibly true. This is also generally supported, but there is no information given on industrial areas. It may not be true of all industrial areas.
(ix) Untrue. Numbers caught at collection points do not differ significantly in industrial areas.
(x) Possibly true. There is no supporting evidence given.

Pupils may need reassuring about the lack of certainty relating to conclusions that can be drawn from the evidence.

b Black moths must have gained a selective advantage during the last century. It is likely that this is related to increased industrialisation and blackening of the environment.

c (i) Black (ii) White

d (i) Black (ii) White

e Black moths are better camouflaged from predators where the tree trunks are dark. The white moths are better camouflaged where the trunks are covered in lichen. Therefore, more of the black forms are eaten in unpolluted woods, but more of the white forms in polluted woods.

f Any reasonable suggestion – e.g. fewer predators; less dense woods so more attracted to the light; less opportunity for dispersal to other woods.

g Under the wings, or out of sight of predators. Otherwise the moths would not be typical of normal moths, and results could be affected by a change in the effectiveness of camouflage.

115 Wheat

Context

Artificial selection. The question relates to some of the practical issues involved in breeding new varieties of wheat for agricultural uses.

Knowledge required

d, **e** Biomass.

f Factors affecting growth of plants.

h to **j** Artificial selection.

Skills

a Data handling (graphs)

b Data interpretation and Reasoning (hypotheses)

c Data interpretation

d Reasoning (application)

e Application (economic)

f Reasoning (application)

g, **h** Data handling (translation)

i Reasoning (explanation)

j Application (technological)

Level

a Mod **b** Mod **c** Mod **d** Mod **e** Mod
f Mod **g** Mod **h** Diff **i** Diff **j** Diff

Answers

a

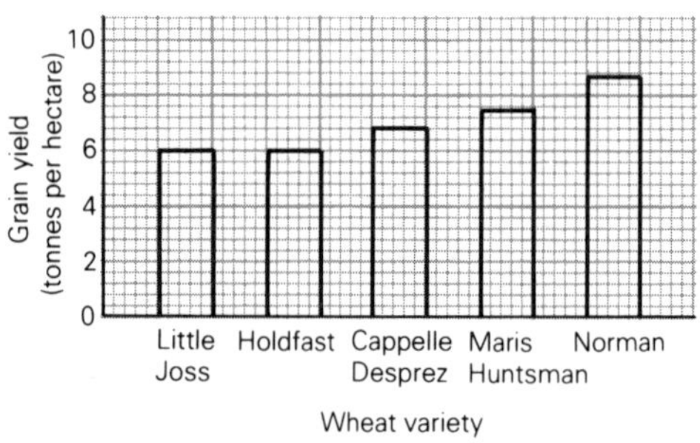

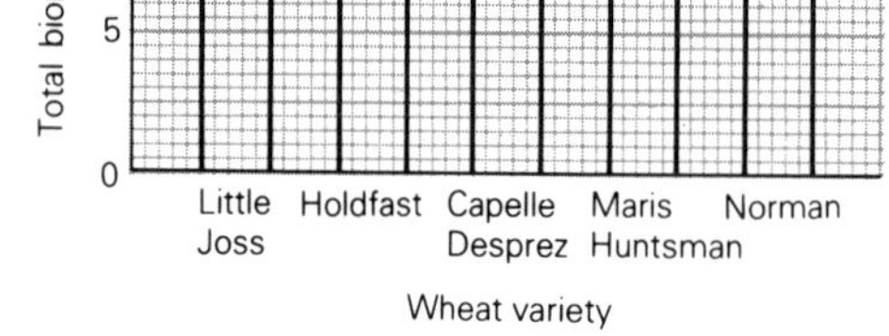

b (i) Height has decreased.
(ii) Shorter stems are less likely to be damaged by wind; less of the nutrients are used to make the stem and therefore more are available to increase grain yield.

c (i) Grain yield has increased (by about 50%).
(ii) There has been very little change in biomass.

d The total mass of the plants.

e The increase in yield has been achieved without a significant increase in biomass. This is a result of the decrease in stem size and means there is a more efficient use of nutrients.

f Mean temperature; availability of water; availability of mineral ions; amount of sunlight; closeness of planting; or any other factor which may affect growth.

g It is more likely to suffer from the rapid spread of a well-adapted pathogen.

h Each new strain will develop diseases caused by pathogens adapted to that particular strain. New strains therefore have to be bred which will not be susceptible to the same pathogens. As the pathogens evolve the cycle has to be repeated.

i The 'seed banks' contain seeds which will have genes not present in the modern varieties. These will enable breeders to select new features, especially disease resistance, which would otherwise not be available.

j Cross the variety with a high yield with a resistant variety; infect the offspring with the fungal disease; select the offspring with most resistance and best yield and breed from these; repeat until a fully resistant variety with a good yield is produced.

116 Adaptation

Context

Pupils are invited to speculate about the origin of various features and forms of behaviour. It is not intended that a 'right answer' should be selected but that plausible hypotheses should be formulated.

Knowledge required

None.

Skills

a to **i** Reasoning (hypotheses)

Level

a Easy+ **b** Easy+ **c** Easy+ **d** Easy+
e Easy+ **f** Easy+ **g** Easy+ **h** Easy+
i Easy+

Answers

Suggestions that have been made, or might be made are:

a The extra pigment in brown eyes gives more protection in bright light.

b The pattern confuses predators by its 'blurring' effect when zebras are running, or it helps herds to keep together. Camouflage seems unlikely as the pattern is very conspicuous on the open plains where zebras often live.

c The red breast acts as a sign stimulus which is used by robins to maintain territory.

d The response may be an advantage in bringing worms to the surface in heavy rain, thus avoiding inundation, although it is not easy to see how this would affect a worm.

e Possibly to flatten vegetation for comfort or to disturb biting insects.

f The white tail acts as a warning signal to other rabbits as it disappears into a burrow, and indicates danger.

g The red spot acts as a stimulus for pecking by young gulls, and this in turn stimulates the parent to open its beak and release food.

h More water would be lost from leaves than could be absorbed from the soil in cold conditions; leaf fall may also reduce wind or snow resistance, or act as a means of removing waste products.

i Flying towards lighter patches may help the moth to move away from concealed hiding places.

First published in Great Britain 1989

British Library Cataloguing in Publication Data
Bailey, Mike
Active Biology.
Teacher's book.
1. Biology – For teaching
I. Title II. Hirst, Keith III. Watkin, Ray
574
ISBN 0 7131 7757 8

Typeset in Rockwell by Tradespools Limited, Frome, Somerset.
Printed in Great Britain for the educational publishing division of Hodder and Stoughton Ltd, Mill Road, Dunton Green, Sevenoaks, Kent by Thomson Litho, East Kilbride, Glasgow.